NEUROPATHOLOGY AND GENETICS OF DEMENTIA

ADVANCES IN EXPERIMENTAL MEDICINE AND BIOLOGY

Recent Volumes in this Series

Volume 477
CELLULAR PEPTIDASES IN IMMUNE FUNCTIONS AND DISEASES 2
Edited by Jürgen Langner and Siegfried Ansorge

Volume 478
SHORT AND LONG TERM EFFECTS OF BREAST FEEDING ON CHILD HEALTH
Edited by Berthold Koletzko, Olle Hernell, and Kim Fleischer Michaelsen

Volume 479
THE BIOLOGY AND PATHOLOGY OF INNATE IMMUNITY MECHANISMS
Edited by Yona Keisari and Itzhak Ofek

Volume 480
BIOLOGY OF THE MAMMARY GLAND
Edited by Jan A. Mol and Roger A. Clegg

Volume 481
ELASTIC FILAMENTS OF THE CELL
Edited by Henk L. Granzier and Gerald H. Pollack

Volume 482
CHROMOGRANINS: Functional and Clinical Aspects
Edited by Karen B. Helle and Dominique Aunis

Volume 483
TAURINE 4: Taurine and Excitable Tissues
Edited by Laura Della Corte, Ryan J. Huxtable, Giampietro Sgaragli, and Keith F. Tipton

Volume 484
PHYLOGENETIC PERSPECTIVES ON THE VERTEBRATE IMMUNE SYSTEM
Edited by Gregory Beck, Manickam Sugumaran, and Edwin L. Cooper

Volume 485
GENES AND PROTEINS UNDERLYING MICROBIAL URINARY TRACT VIRULENCE: Basic Aspects and Applications
Edited by Levente Emődy, Tibor Pál, Jörg Hacker, and Gabriele Blum-Oehler

Volume 486
PURINE AND PYRIMIDINE METABOLISM IN MAN X
Edited by Esther Zoref-Shani and Oded Sperling

Volume 487
NEUROPATHOLOGY AND GENETICS OF DEMENTIA
Edited by Markus Tolnay and Alphonse Probst

NEUROPATHOLOGY AND GENETICS OF DEMENTIA

Edited by

MARKUS TOLNAY

and

ALPHONSE PROBST

University of Basel
Basel, Switzerland

SPRINGER SCIENCE+BUSINESS MEDIA, LLC

Library of Congress Cataloging-in-Publication Data

Neuropathology and genetics of dementia/edited by Markus Tolnay and Alphonse Probst.
p. cm. — (Advances in experimental medicine and biology; v. 487)
ISBN 978-0-306-46558-1 ISBN 978-1-4615-1249-3 (eBook)
DOI 10.1007/978-1-4615-1249-3
1. Dementia—Pathophysiology—Congresses. 2. Dementia—Genetic aspects—Congresses. 3. Nervous system—Pathophysiology—Congresses. I. Tolnay, Markus, 1962- II. Probst, Alphonse, 1938— III. Series.

RC521 .N47 2001
616.8'3—dc21

00-067413

Proceedings of the Swiss Society for Neuropathology XVIIIth International Winter Meeting on Neuropathology and Genetics of Dementia, held March 23–26, 2000, in St. Moritz, Switzerland

ISSN: 0065 2598

ISBN 978-0-306-46558-1

Originally published by Kluwer Academic/Plenum Publishers, New York in 2001

10 9 8 7 6 5 4 3 2 1

A C.I.P. record for this book is available from the Library of Congress

Preface

For more than 35 years the Swiss Society of Neuropathology has organised its traditional International Winter Meeting. The main aim of this venue is to bring together neuropathologists and clinicians as well as neuroscientists interested in disease mechanisms. The XVIIIth International Winter Meeting was held in March 2000 in St. Moritz, Switzerland. The topic of the 2000 Meeting was "Neuropathology and Genetics of Dementia". A programme of invited plenary lectures of high educational value as well as platform and poster presentations given by many participants covered the broad spectrum of dementing disorders. Encouraged by the high standard of the meeting, and probably also influenced by the advent of a new Millennium we decided, together with Kluwer Academic/Plenum Publishers, to publish the Proceedings of the 2000 Meeting in the present book.

The increase of the aged population challenges social and health care systems in industrialized societies. Maintenance of health and autonomy into old age is a key factor, and the autonomy and independence of old persons crucially depend on mental health. However, neurodegenerative disorders accompanied by dementia affect 5-10% of individuals over the age of 65, and for the age group between 65 and 95 the prevalence of cognitive dysfunction doubles every 5 years. With this challenge in mind, Rossor discusses the clinical syndrome of dementia which encompasses an enormous variety of underlying disorders.

Over the last few years it has become apparent that many neurodegenerative disorders are characterised by protein misfolding and aberrant polymerization and may therefore be viewed as "aggregation proteopathies". Among them a substantial subset of dementing disorders, either with or without motor dysfunction, are characterised by abundant neuronal and/or glial accumulation of hyperphosphorylated tau protein. The molecular parameters of these "tauopathies" are reviewed by Delacourte. The recent discovery that mutations of the tau gene are causative for familial forms of frontotemporal dementia and Parkinsonism (FTDP-17) brought the "tau pathway" and the "tauopathies" to the centre stage. This extraordinary breakthrough will be covered by Spillantini and Goedert. There is increasing evidence that argyrophilic grain disease is much more frequent in older patients than previously thought. Tolnay et al. are reviewing recent contributions to the pathology of this late-onset dementing disorder in which tau pathology, in the almost total absence of amyloid beta deposits, is almost exclusively restricted to the limbic lobe and the amygdala. Schultz et al. report an intriguing age-related progression of neuronal and glial tau pathology in baboons thus rendering the baboon a potential primate model for age-related human „tauopathies". Götz et al. introduce transgenic mice expressing mutated human tau as valuable animal models for reproducing formation of neuro- and glial fibrillary lesions.

AD is neuropathologically characterised by abundant extracellular amyloid beta deposits (among them the neuritic plaques) and intracellular neurofibrillary tangles. Gandy and Petanceska are reviewing the hitherto proposed mechanisms of amyloid precursor protein (APP) trafficking and metabolism. In addition they focus on the recent findings that estradiol might influence the risk or age-at-onset of AD by lowering the speed of the accumulation of amyloid beta protein. The role of type 10 17beta-hydroxysteroid dehydrogenase in this process is reviewed by Yang and He. In AD but also in non-demented elderly subjects the amyloid beta protein accumulates in the walls of small cerebral blood vessels (cerebral amyloid angiopathy; CAA). Weller and Preston present evidence for the hypothesis that CAA results from the accumulation of amyloid peptides in perivascular interstitial fluid drainage pathways and that the impedence of drainage of amyloid beta protein from the brain may contribute to the onset of AD. The contribution by Herzig et al. on APP transgenic mice which develop abundant CAA is in line with this hypothesis.

Lewy bodies and Lewy neurites are hallmark lesions of Parkinson's disease (PD) and Dementia with Lewy bodies (DLB), and the main protein constituent of these lesions is the presynaptic protein alpha-synuclein. Alpha-synuclein is also present in glial cytoplasmic inclusions of multiple system atrophy (MSA), and therefore PD, DLB and MSA have been lumped under the term "alpha-synucleinopathies". Jensen and Gai review the role of alpha-synuclein in neurodegeneration while the contribution by Ince focuses on the relationship between AD, PD and DLB at the clinicopathological level. Recent findings *in vivo* in transgenic flies and transgenic mice provide novel perspectives to model in animals the pathophysiology of diseases with alpha-synucleinopathy. These new aspects are covered by the contribution by Barbieri et al.

A large number of disease-related cellular features can be detected by ubiquitin immunohistochemistry. While for some disorders the molecular nature of the ubiquitinated filamentous inclusions is well known, e.g. amyloid beta protein, tau protein, alpha-synuclein and huntingtin, there is a substantial subset of neurodegenerative disorders for which the nature of the pathological protein complexed with ubiquitin still remains unknown. Among others, motor neuron disease, frontotemporal dementia with inclusions of motor neuron disease type and some cases of primary progressive aphasia and semantic dementia belong to this group of „ubiquitin disorders". The involvement of the ubiquitin system in diseases of the nervous system will be covered by Lowe et al. Furthermore, using ubiquitin-immunohistochemistry, Probst et al. describe a novel type of hippocampal ubiquitin pathology in AD and Pick's disease.

Polyglutamine expansion diseases such as Huntington's disease, spinobulbar muscular atrophy and spinocerebellar ataxia constitute a subset of "proteopathies" in some of which there are abundant neuronal intranuclear polyglutamine inclusions. The pathological mechanisms underlying these disorders will be reviewed by Devys et al.

Though most of us would associate the term dementia with old age, dementia may also occur in the pediatric age group in a variety of progressive neurological disorders. The contribution by Goebel and Kohlschütter deals with the presence of dementia in the neuronal ceroid-lipofuscinoses.

The reader might ask why inclusion body myositis (IBM) is included in a book devoted to dementia. However, it has become apparent in the last few years that IBM inclusions encompass a wide range of proteins involved in neurodegenerative dementing disorders, such as APP, tau, presenilin-1, apolipoprotein E and prion protein. Lampe et al. are focusing on the role of neurodegeneration-associated proteins in IBM.

Last, but not least, the book closes with the contribution by Turner and Schapira who critically review the evidence for the role of mitochondrial dysfunction in a wide range of neurodegenerative disorders and the ageing brain.

Together with the contributors we believe that this book will not only be useful to neuropathologists but also to clinicians and basic scientists interested in the broad field of dementia.

Markus Tolnay
Alphonse Probst

Acknowledgments

We thank the authors for their contributions, the publishers, particularly Joanna Lawrence, for their help in the production of this book and Thomas Schürch and Michelle Pfeiffer for their photographic and administrative help. Finally, we are grateful to Kurt Brauchli who guided us safely through all expected, but even more important, unexpected computer problems.

Markus Tolnay
Alphonse Probst

Contents

CAN THE PHYSICIAN PREDICT THE NEUROPATHOLOGIST?

Martin N. Rossor
The National Hospital for Neurology and Neurosurgery, Queen Square, London WC1N 3BG, UK.

The clinical syndrome of dementia encompasses an enormous variety of underlying diseases. Some present as relatively consistent clinicopathological entities, whereas with others the relationship between the clinical syndrome and the underlying molecular pathology can be extremely variable. In general, the most robust associations are between the clinical syndrome and the area of cerebral cortex or subcortical structures involved. The latter may relate to the underlying molecular pathology by virtue of selective vulnerability of particular neuronal systems to a pathological process but not necessarily so.

The term dementia is used to describe patients with multiple domains of cognitive impairment of which one must be memory. The cognitive impairment must also be sufficiently severe to interfere with employment and/or social functioning. The clinical concept of dementia was invaluable in the pre-scanning days of clinical neurology, as it served to distinguish the patient with a focal lesion for example a meningioma, from a patient with a more widespread degenerative process. It is now less useful especially since any disease process will be well established by the time the dementia criteria are met, i.e. memory impairment plus at least one other cognitive deficit. Thus patients with Alzheimer's disease (AD) presenting with a memory impairment, may have a prolonged period before other cognitive domains such as language or visuoperceptual function become affected. Strictly speaking, these patients cannot be diagnosed clinically as having AD because they do not fulfil criteria for dementia, a *sine qua non* for the

Neuropathology and Genetics of Dementia, Edited by Tolnay and Probst
Kluwer Academic/Plenum Publishers, New York 2001

clinical criteria of AD to be fulfilled. Similarly, patients with Pick's disease, a prototypical degenerative dementing disease, may only have language impairment for many years.

The causes of dementia are legion. Many diseases may result in cognitive impairment as part of a more generalised neurological disturbance. Often the cognitive dysfunction is only one part of a disease better characterised by a cerebellar ataxia or spastic paraparesis. These "dementia plus" syndromes are usually characterised by the non-cognitive features. Of the primary degenerative dementias, AD is by far the commonest cause, either alone or in combination with a variable degree of vascular disease. The typical presentation of AD is with an impairment of episodic or event memory i.e. the memory for day to day events which informs our autobiographical experience. Using the National Institute for Neurological and Communicative Diseases - Alzheimer's Disease and Related Disorders Association (NINCDS-ADRDA) criteria, approximately 80% of patients are successfully diagnosed in specialist centres. The diagnostic accuracy is less in younger patients, reflecting the increased prevalence of non-AD dementias in this younger age group. Less common presentations of AD include frontal dysexecutive syndromes, dysphasias and visual disorientation. The latter is a striking clinical syndrome in which the localisation of objects in space, an aspect of early visual processing, is impaired. Such patients have been subsumed within the rubric of posterior cortical atrophy[1] and the majority turn out to have AD. Their visual failure can appear inconsistent to the unwary clinician and is often dismissed.

The greatest diagnostic challenge lies with the group of disorders that are associated with selective frontotemporal lobar atrophy. In general, the clinical syndrome reflects the topography of the neurodegeneration rather than the underlying molecular pathology. This can include a variety of tauopathies such as Pick's disease, corticobasal degeneration and familial dementia associated with mutations in the tau gene; tau-negative ubiquitin-positive inclusion body degeneration and non-specific degenerations without specific features. In addition, Creutzfeldt-Jakob disease and rarely AD, can also present as a frontotemporal dementia.

Three prototypic clinical syndromes have been described with frontotemporal degeneration (FTD): frontotemporal dementia, progressive non-fluent aphasia and semantic dementia[2]. Frontotemporal dementia describes patients with behavioural disturbances and predominantly a frontal dysexecutive syndrome. Progressive non-fluent aphasia is an uncommon presentation that was originally described as primary progressive aphasia by

Mesulam[3]. Patients with semantic dementia by contrast to those with non-fluent aphasia, have a strikingly fluent but empty speech arising from impairment in semantic verbal memory. The term semantic dementia relies upon the distinction drawn by Tulving[4] between our episodic or event memory and semantic memory, a culturally based knowledge of meaning. This involves both meaning for words, our verbal semantic memory and for objects, our visual semantic memory. Warrington described the first case of selective impairment in verbal semantic memory in a patient with a frontotemporal degeneration[5,6] and the term semantic dementia was subsequently introduced by Snowden[7]. In many patients, the semantic memory impairment will extend from the verbal to the visual domain with the development of a visual associative agnosia. Most, but not all cases of FTD are asymmetric and some strikingly so. The majority are left-sided and associated with semantic dementia or progressive non-fluent aphasia. More rarely there may be selective right-sided frontotemporal atrophy. In addition to behavioural disturbance, such patients may present with a prosopagnosia or inability to identify and recognise faces[8,9].

Can the physician predict the neuropathologist? Clearly prediction of a specific underlying molecular pathology is unreliable. The relevance of a diagnosis however, depends on the question being asked of the clinician. In the clinical management of cognitive and functional impairment, an understanding of the underlying disease process may not be critical. However, as we move toward specific therapeutic interventions, a specific molecular diagnosis in vivo will become more important; clearly, clinical descriptions are inadequate for this task. A specific molecular diagnosis is likely to need diagnostic adjuncts such as specific imaging e.g. of amyloid or tau deposition and CSF biochemical markers. These are all being actively pursued but as yet are not ready for the clinic.

REFERENCES

1. Benson F, Davis J, Snyder BD (1988) Posterior cortical atrophy. Arch Neurol 45: 789-793.
2. Neary D, Snowden JS, Gustafson L et al. (1998) Frontotemporal lobar degeneration: a consensus on clinical diagnostic criteria. Neurology 51: 1546-1554.
3. Mesulam MM (1982) Slowly progressive aphasia without generalized dementia. Ann Neurol 11: 592-598.
4. Tulving E (1972) Episodic and semantic memory. In: Organization of memory (Tulving E, Donaldson W eds), pp. 382-403. Academic Press, NY, London.
5. Warrington EK (1975) The selective impairment of semantic memory. Q J Exp Psychol 27: 635-657.

6. Rossor MN, Revesz T, Lantos PL et al. (2000) Semantic dementia with ubiquitin-positive tau-negative inclusion bodies. Brain 123: 267-276.
7. Snowden JS, Goulding PJ, Neary D (1989) Semantic dementia: a form of circumscribed cerebral atrophy. Behav Neurol 2: 167-182.
8. Evans JJ, Heggs AJ, Antoun N et al. (1995) Progressive prosopagnosia associated with selective right temporal lobe atrophy. A new syndrome? Brain 118: 1-13.
9. Tyrrell PJ, Warrington EK, Frackowiak RS et al. (1990) Progressive degeneration of the right temporal lobe studied with positron emission tomography. J Neurol Neurosurg Psychiatry 53: 1046-1050.

THE MOLECULAR PARAMETERS OF TAU PATHOLOGY

Tau as a killer and a witness

André Delacourte
Inserm U422, Place de Verdun, 59045 Lille cedex

1. INTRODUCTION

Tau proteins are markers of different pathological conditions, such as hypoxia, different types of stress and degenerative processes observed in nerve cells, glial cells or muscle fibers. Tau proteins are the basic components of the pathological filaments that accumulate in neurons and glial cells affected by neurofibrillary degeneration[1]. Tau pathology is observed in more than 20 different diseases, including Alzheimer's disease (AD), progressive supranuclear palsy (PSP), corticobasal degeneration (CBD), Pick's disease (PiD), myotonic dystrophy (MyoD), familial frontotemporal dementias with Parkinsonism associated to chromosome 17 (FTDP-17), etc. Tau is an outstanding marker well correlated with clinical manifestations. Indeed, pathological tau proteins observed in the association neocortical areas are systematically associated with dementia. Together, tau pathology reveals precisely the intensity and the extent of the degenerating process. This pathological entity can be defined according to 6 different molecular parameters that are presented here.

Neuropathology and Genetics of Dementia, Edited by Tolnay and Probst
Kluwer Academic/Plenum Publishers, New York 2001

2. NEUROFIBRILLARY DEGENERATION (NFD)

2.1 A modern definition

Neurofibrillary degeneration (NFD) is a degenerating process visualized at the histological level by silver stains, as demonstrated by Aloïs Alzheimer and colleagues at the beginning of the 20th century. This technique is still used for neuropathological examination, and reveals in detail the abnormal intracellular fibrils that accumulate in cell bodies and in neurites: neurofibrillary tangles, neuropil threads as well as the wreath of dystrophic neurites of senile plaques. These lesions characterize Alzheimer's disease. NFD is also observed in numerous other degenerative disorders. All approaches combined, the modern definition of neurofibrillary degeneration is the following: a degenerating process characterized by the abnormal filamentous accumulation of tau proteins in neurons and glial cells.

2.2 Pathologies with neurofibrillary degeneration

NFD is observed in more than 20 other neurodegenerative disorders. All these diseases are very different, in that they are familial or sporadic, with different origins, from traumatism to a mutation on tau gene. NFD can affect preferentially subcortical nuclei or neocortical areas, neurons or in addition astrocytes and/or oligodendrocytes. The pattern of fibrils is also different and characteristic, such as the Pick bodies of Pick's disease (PiD) or the neuritic plaques of Alzheimer's disease. At the electron microscopic level, the filamentous material of NFD is either helical, twisted or straight, according to the neurodegenerative disorders[2].

2.3 The normal role of tau proteins

Tau proteins belong to the microtubule-associated proteins (MAP) family[1]. The human tau gene is unique and located over 100 kb on the long arm of chromosome 17 at band position 17q21, and contains 16 exons. Exons 2, 3 and 10 are alternatively spliced and are adult brain-specific. In the human brain, the tau primary transcript gives rise to six mRNAs, three of them with exon 10. Translation of exon 10 adds a fourth repeated sequence (4R) which is a binding site to tubulin dimers, the basic components of microtubules. The normal role of tau is to stabilize microtubules, which are the tracks of the intraneuronal transport. Stabilization of microtubules is dramatically increased by tau isoforms with 4 repeated binding sites (4R tau or tau E10+ isoforms) (Figure 1). Conversely, phosphorylation of tau destabilizes microtubules and it is suggested that abnormal phosphorylation, as observed in AD, provokes a collapse of the microtubule network.

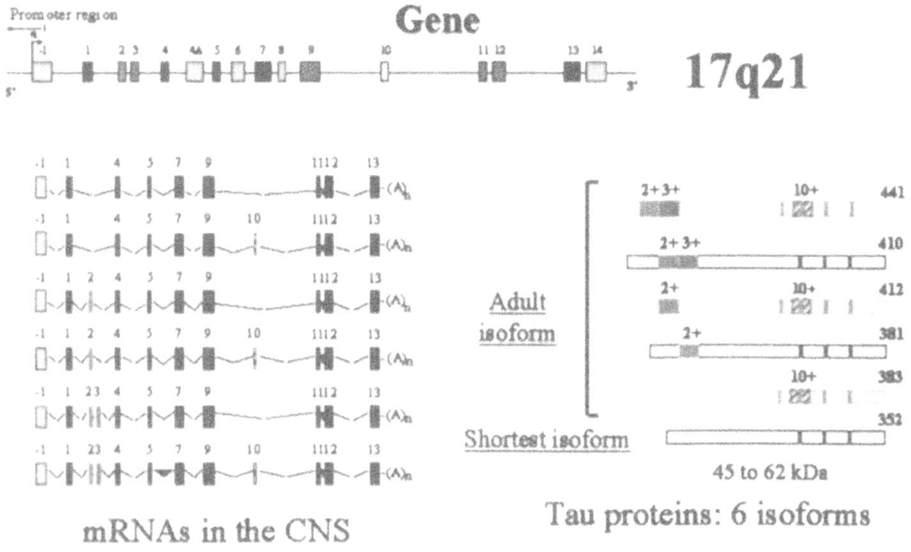

Figure 1. Expression of 6 tau isoforms in the human central nervous system

3. THE PARAMETERS OF TAU PATHOLOGY IN THE HUMAN BRAIN

3.1 Quantification of abnormal tau proteins

Antibodies against phosphorylation sites on tau, such as AD2 directed against phosphorylated serines 396 and 404, are able to specifically detect pathological tau in AD[3].

As shown in Figure 2, a characteristic triplet of pathological tau proteins is exclusively detected in Alzheimer brain extracts from polymodal association brain areas while no trace of tau pathology is observed in the same brain areas from non-demented aged-matched controls. Pathological tau proteins from AD, named PHF-tau, are composed of three main electrophoretic variants designed as Tau 60, 64, 69, as a function of their molecular mass. A minor fourth band named Tau 74 is also detected at 74 kDa[4,5]. Furthermore, the intensity of the detection is proportional to the intensity of NFD. Therefore, a semi-quantification by Western blot is able to easily detect and quantify NFD, in good agreement with immunohistochemical observations. However, we have observed in some cases that the detection of pathological tau was detected biochemically before the presence of tangles at the immunohistological level.

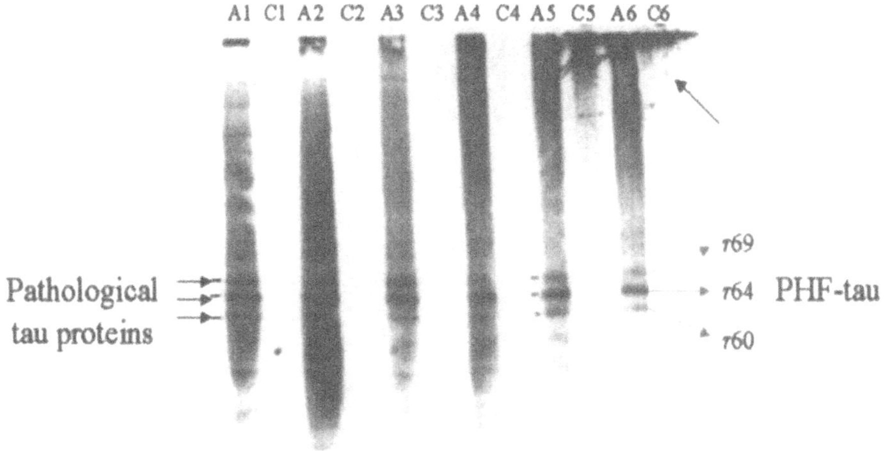

A1-A6 : Alzheimer's disease parietal cortex samples, C1-C6 : Control cortex samples

Figure 2. Immunoblot detection of pathological tau protein in Alzheimer brain extracts, with AD2, a phospho-dependent tau antibody, as described by Sergeant et al.[4]. Same amounts of total protein extracts were loaded in each well. The same brain area was analyzed, the parietal cortex. Controls were non-demented aged-matched cases. Pathological tau proteins are detected, with a molecular mass of 60, 64 and 69 kDa. Pathological tau proteins are exclusively detected in Alzheimer brain extracts. Note the intensity of tau pathology which is different among the Alzheimer patients.

3.2 Different biochemical signatures in AD, PSP, PiD and MyoD

As shown in the previous section, a characteristic tau triplet is detected in AD brain homogenates. A different, but also characteristic pattern, is observed in other diseases with pathological tau and with tau positive-brain lesions.

Indeed, we demonstrated that pathological tau proteins from progressive supranuclear palsy (PSP) are composed of a main doublet (tau 64, tau 69)[6] and a minor tau 74[4], while those from Pick's disease are composed of another doublet (tau 60, 64) and a minor tau 69[7].

Myotonic dystrophy (MyoD), a familial disease with abnormal CTG repeats on chromosome 19, is characterized by a tau pathology with a major band of 60 kDa. Electrophoretic bands at 64 and 69 are also found, but in lower amounts[8] (Figure 3).

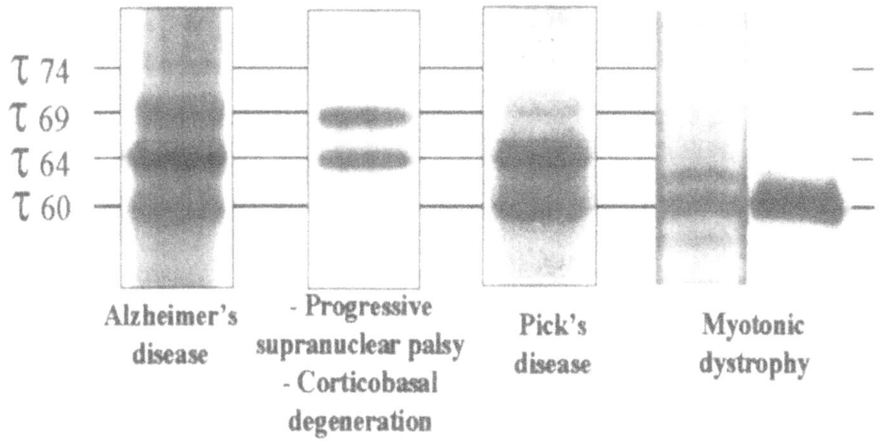

Figure 3. The different tau signatures in several disorders.

3.3 Tau isoforms in brain lesions: class I to IV

The analysis of the tau isoform content, using specific immunoprobes against isoforms with exon 2, 3 or 10, and 2-dimensional gels, enables us to characterize the different patterns of pathological tau. Specific sets of tau isoforms aggregate to constitute 4 main classes of brain lesions: AD (all 6 isoforms), PSP/CBD (3 E10+ isoforms)[4], PiD (3 E10- isoforms)[7] and MyoD (mainly the shortest tau)[8]. These different patterns can be reconstructed in cellular models[9].

These different classes of tau isoform patterns can be explained at two levels: the first one is that neuronal subsets could express specific sets of tau isoforms. For example, it has been shown that the granule cells of the dentate gyrus do not normally express E10+ tau isoforms. If a given neuronal subset degenerate, the corresponding set of tau isoforms of this neuronal subset will aggregate to constitute a characteristic tau lesion. In our example, neurons that express only E10- tau isoforms will form a Pick body (Figure 4).

On the other hand, E10+ tau isoforms are able to dramatically increase the stability of microtubules and therefore to modify the physiological properties of the cell. It is likely that the expression of E10+ tau isoforms can be dysregulated in many diseases, due to an abnormal alternative splicing. Overexpression of E10+ isoforms is demonstrated in FTDP-17, following a pathogenic mutation in the intron 10 regulating the splicing of exon 10 (reviewed in[2]). This abnormal processing of intron 10 could also be boosted by polymorphisms such as those found in PSP[10].

Together, we observe an important number of neurodegenerative diseases, with different etiologies and classes of tau pathology (Table 1). They have all in common a cognitive impairment when tau pathology is present in association polymodal brain areas. For some of them, tau is the etiological agent (FTDP-17). For many other diseases tau pathology is not only a marker, but also a motor of the degenerating process (AD, PSP, CBD)[2]. For dementia pugilistica, tau pathology is a consequence of the repeated traumatisms.

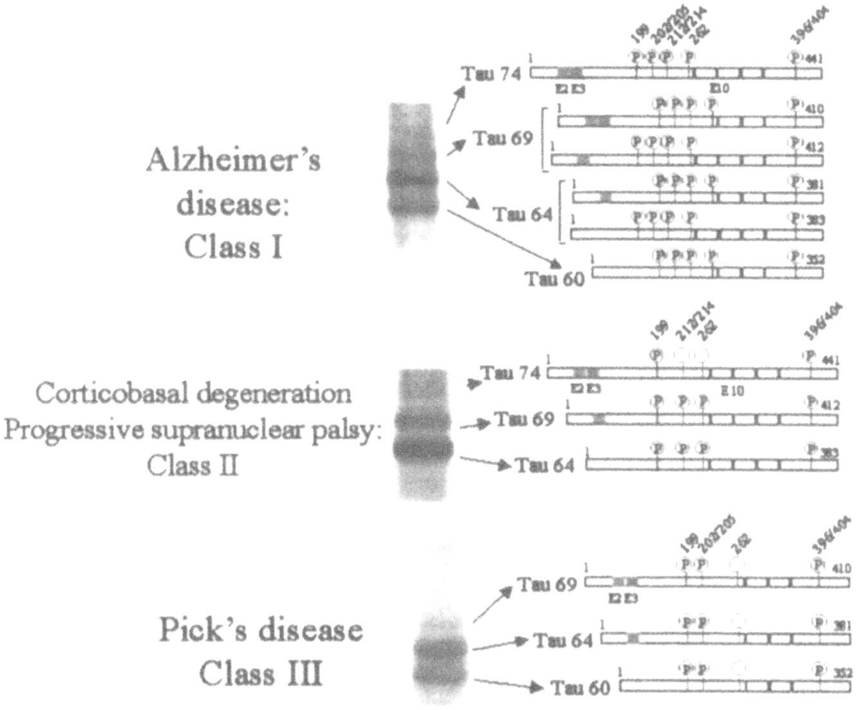

Figure 4. The correspondence between the biochemical signature of tau pathology and the tau isoform content

Disease	Classes of tau pathology
Aging (hippocampal region, patients over 75 years)	I
Alzheimer's disease, familial and sporadic	I
Amyotrophic lateral sclerosis/parkinsonism-dementia complex of Guam	I
Argyrophilic grain dementia	
British type amyloid angiopathy	I
Corticobasal degeneration	II
Dementia pugilistica/autism with self-injury behaviour	I
Down's syndrome	I
FTDP-17	II, I and III
Gerstmann-Straussler-Scheinker disease (rarely)	I
Hallenvorden-Spatz disease	
Inclusion body myositis	
Multiple system atrophy	
Myotonic dystrophy	IV
Niemann-Pick disease type C	I
Pick's disease	III
Presenile dementia with tangles and calcifications	
Prion protein cerebral amyloid angiopathy	
Progressive supranuclear palsy	II
Post-encephalitic parkinsonism	I
Subacute sclerosing panencephalitis	
Tangle only dementia	

Table I. Presentation of the different neurodegenerative disorders with a tau pathology, and their different biochemical tau signatures, from Class I to Class IV.

3.4 Disease-specific sites of phosphorylation

3.4.1 Native tau are phosphorylated

Native tau proteins are normally phosphorylated on numerous serine or threonine sites (80 potential sites). In a living cell, there are different pools of tau proteins that are more or less phosphorylated. Tau proteins with high phosphorylation levels are not bound to microtubules. The state of phosphorylation of tau proteins is likely different according to the cell compartments. Thus, tau are less phosphorylated in axons, as demonstrated by monoclonal antibody tau1. In addition, phosphorylation of tau proteins is developmentally regulated. The antibodies such as AT8, AD2, PHF-1, that are well known in the Alzheimer field to label tau pathology, also strongly label native tau proteins (reviewed in[1]). Tau proteins phosphorylated at the 396/404 site (AD2 site) are free proteins not bound to tubulin[11].

Alzheimer's biochemistry is always performed on post-mortem human brains. Following the pioneer work of Matsuo et al.[12], we know that native tau proteins are almost totally dephosphorylated during post-mortem delays.

The dephosphorylation results from the strong phosphatasic activity which is released after cell death.. On the other hand, aggregated tau proteins that constitute brain lesions are not dephosphorylated, because the phosphatase enzymes are unable to access phosphorylated sites that are buried deep in brain lesions. Therefore, using antibodies such as AT8 or AD2 on the post-mortem tissue, normal tau proteins will not be detected, because they are dephosphorylated. Conversely, tau proteins inserted in brain lesions, that are still phosphorylated, will be detected by AT8 or AD2 (Figure 5).

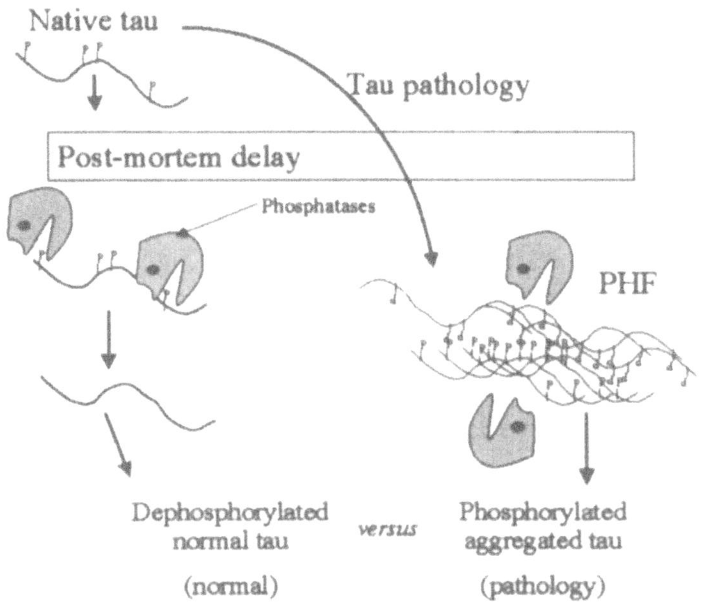

Figure 5. Cartoon explaining why a phospho-dependent anti-tau antibody reveals an aggregation of tau.

Conclusion: Western blots such as the one presented in Figure 2 reveal aggregated tau proteins and tau pathology. This immunochemical labeling, carried out with a phospho-dependent antibody that detects native tau, reveals the aggregation of tau, but not its hyperphosphorylation, as frequently mentioned in the literature (Figure 5).

3.4.2 Tau proteins in Alzheimer brains are hyperphosphorylated

The previous section explained why a Western blot analysis of a biopsy sample from a human brain reveals a tau triplet which is similar to the triplet of Alzheimer's disease. The major difference is that no smears of aggregated tau are observed in the normal brain tissue. Therefore, the question is to determine if tau proteins in Alzheimer brain are really hyperphosphorylated, as frequently suggested[1].

Sergeant et al.[13] have been able to directly demonstrate the hyperphosphorylation of tau, by comparing on 2D gel the isoelectric pattern of tau proteins from a biopsy of a normal human brain tissue versus an autopsy sample of an Alzheimer brain. The 2D electrophoresis revealed that Alzheimer tau are more acidic, and therefore more phosphorylated, than normal native tau. These results demonstrate that pathological tau from AD are hyperphosphorylated.

3.4.3 Tau proteins in Alzheimer brains are abnormally phosphorylated

A few sites on tau proteins have been shown to be phosphorylated on PHF-tau, but not present on native tau. These sites are Ser 212/Ser214 detected by AT100, Ser 231/Ser235 detected by PHF-27 and TG3, and Ser 422 detected by AP422 or Ab988 (reviewed in[1]). These monoclonal antibodies demonstrate the aberrant phosphorylation during the degenerating process (Figure 6).

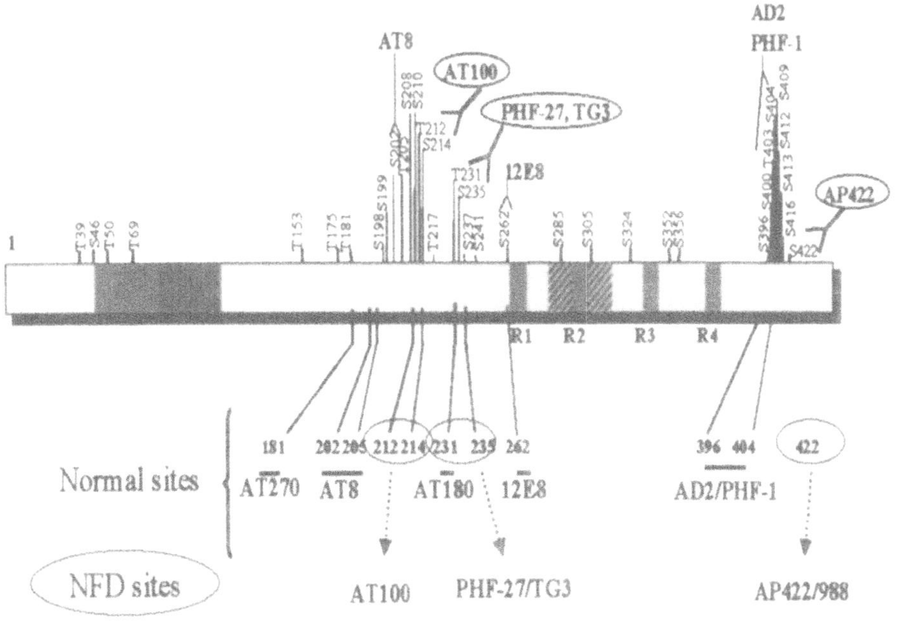

Figure 6. The different sites of phosphorylation, normal or pathological, in diseases affected by neurofibrillary degeneration

3.4.4 Some phosphorylation sites are disease-specific

Many normal phosphorylation sites, such as AT8 or AD2, are observed in all tau lesions of the neurodegenerative diseases presented in Table 1. In the same way, phosphorylated pathological epitopes such as Ser422 are found in all tau lesions, whatever the pathology[14]. However, there are some interesting specificities. Indeed, it has been shown at the immunohistological and the biochemical levels that Ser 262 is phosphorylated in many tau lesions, but not in Pick bodies of PiD[7,15]. Also, tau inclusions in oligodendrocytes of MSA do not contain pathological sites of phosphorylation[16].

3.5 Spatio-temporal distribution

The mapping of the spatial-temporal distribution of tau pathology in the different brain areas is important to understand how the disease spreads in the brain. Indeed, there is a precise pathway of tau pathology in ageing and in AD. The progression of tau pathology is sequential, invariable, hierarchical, and predictable. Ten stages (S1 to S10) were defined, corresponding to 10 brain areas sequentially affected. The degenerating process always starts in the hippocampal region (S1: transentorhinal cortex; S2: entorhinal cortex; S3: CA1 region of the hippocampus), followed by the temporal cortex (S4: temporal pole; S5: inferior temporal cortex; S6: mid temporal cortex), then the polymodal association brain areas (S7-S8), and finally the primary regions (motor cortex and/or the occipital cortex) as well as many subcortical nuclei (S9-S10)[17].

These data show that tau pathology is systematically observed in people aged over 75 years, sometimes independently of amyloid deposits and that the hippocampal area is the most vulnerable area of the human brain. This hippocampal vulnerability is probably a springboard for Alzheimer pathology, namely amyloid precursor protein dysfunction, which will exacerbate and extend tau pathology in other brain areas. Together, neuropathological and biochemical data concur to indicate that tau pathology is instrumental in AD.

From these data, it is now possible to precisely quantify tau pathology and amyloid deposition at the biochemical level and to determine the Criteria to Establish a Biochemical Diagnosis of AD (CEBDAD), that separate ageing from infraclinical AD, and infraclinical AD from clinical AD[17]. In that respect, it should be pointed out that recent criteria for the neuropathological diagnosis have rehabilitated tau pathology, in good agreement with Aloïs Alzheimer's observations and the natural history of AD[18].

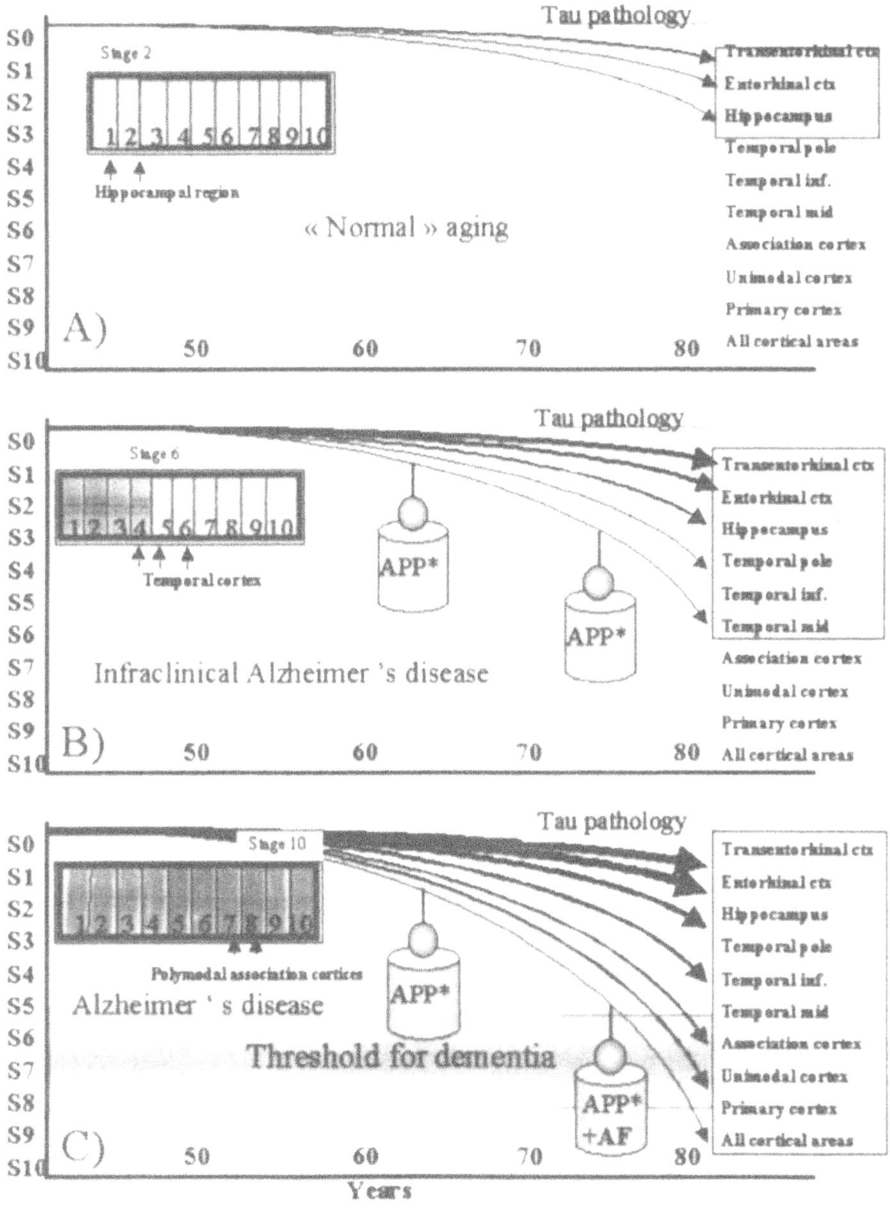

Figure 7. The biochemical pathway of tau pathology in ageing and Alzheimer's disease. Insert: Western blot analysis in the 10 brain areas successively affected in AD. **Facts:** Tau pathology is systematically observed in the transentorhinal cortex (stage 1), or in addition in the entorhinal cortex (stage 2), and CA1 of hippocampus (stage 3) in non-demented patients aged over 75 years. Tau pathology can be found in other brain areas, along a stereotypical pathway. All brain areas are affected by tau pathology in Alzheimer's disease. **Hypothesis:** the spreading of tau pathology is fuelled by APP dysfunctions.

The pathway of tau pathology in PSP is also stereotypical, but different from the one in AD, and roughly the opposite, emerging from subcortical nuclei to the neocortex. The specific pattern of PSP, with the upper tau doublet (tau 64, tau 69) is observed in all brain areas affected, from subcortical nuclei to frontal neocortical regions[19, 20].

In all types of neurodegenerative disorders, the spreading of tau pathology follows specific neuronal connections, like a precise neuronal chain reaction. There is also likely a "dynamic" of spreading, fuelled by different factors, each of them being a lead for neuroprotection[21].

3.6 The genetic profile of tau pathology

3.6.1 Tau pathology as an aetiological agent

The presence of brain lesions in neurodegenerative disorders usually arouses the same question: is this a cause or a consequence? Some answers are already available for tau pathology.

3.6.1.1 Frontotemporal dementia with parkinsonism linked to chromosome 17 (FTDP-17)

The discovery that tau mutations are directly involved in numerous FTDP-17 has been dramatically documented. More than 20 different mutations have been spotted. Most of the pathogenic mutations are responsible for an increase of 4R tau isoforms, giving a class II tau pathology. 4R tau isoforms, with the additional peptidic sequence of exon 10, have a much stronger affinity towards microtubules than 3R isoforms, and an excess could abnormally modify microtubules, which will be stiffer and less dynamic. The other mutations are missense mutations that also affect microtubule polymerization and stability by decreasing tau–microtubule binding and lead to one of the class I, II or III tau pathologies. Interestingly, some of the mutations present in all tau isoforms, such as R406W, are responsible for a specific loss of functions of 3R tau isoforms[22].

The striking feature of these familial tauopathies is the heterogeneity of the phenotype which results from the different effects on tau (overexpression, loss of function). Surprisingly, for the same mutation in the same family, different onsets and different phenotypes can be observed, showing that numerous additional factors can bring even more heterogeneity to the phenotype (reviewed in[2]).

We also noted that the neuropathological profile in FTDP-17 is quite different from AD, with a special involvement of astrocytes and cortical white matter, but also with an important heterogeneity for each mutation.

However, not all diseases with a tau pathology have mutations on the tau gene. This has been verified for CBD, PSP, amyotrophic lateral sclerosis/parkinsonism dementia complex of Guam, and AD.

3.6.1.2 Polymorphisms in progressive supranuclear palsy

Conrad and colleagues[23] identified a polymorphic dinucleotide repeat sequence in a Caucasian population with PSP. This polymorphism, named A0, corresponding to an 11 TG dinucleotide repeat in intron 9 of the tau gene, is found in 95% of the PSP cohort (95.5%) and only in 57% of normal controls and 50% of patients with AD. Recently, these data were confirmed by several studies and extended to a haplotype including a number of polymorphisms in linkage disequilibrium with A0 and named H1. This haplotype corresponds to A0 polymorphism, numerous single nucleotide polymorphisms along the entire tau gene and one intronic 238 bp deletion flanking exon 10. These polymorphisms may influence exon 10 splicing and thus the proportion of 4R:3R tau isoforms, leading to a class II tau pathology. It should be noted that these A0 polymorphisms or H1 haplotypes were recently described in other pathologies including CBD and Parkinson's disease. Some other polymorphisms in the tau gene were also described as being associated with a risk of AD, but these data are still controversial (reviewed in[2]).

3.7 Conclusion

Together, the correlation between the distribution of tau pathology and clinical manifestation is excellent if we take into account the extent and the function of the brain areas that are affected. In AD, dementia is observed at stage 7 or above, when polymodal association brain areas are affected. In PSP and CBD, dementia is always observed when tau pathology is found in frontal neocortical areas.

Other parameters of tau pathology might be important, such as the extent of tau phosphorylation, and the location of the phosphorylated sites. Other post-translational events such as glycosylation, that modulate tau functions or sorting in cell compartments have to be investigated[1]. Tau pathologies are also influenced at the genetic level, either by polymorphisms or directly driven by mutations on tau gene. In conclusion, tau pathology is a many-sided degenerating process that will open diagnostic and therapeutic avenues.

REFERENCES

1. Buée L, Bussière T, Buée-Scherrer V et al. (2000) Tau protein isoforms, phosphorylation and role in neurodegenerative disorders. Brain Res Rev: in press.
2. Delacourte A, Buée L (2000) Tau pathology: a marker of neurodegenerative disorders. Current Opinion in Neurology: in press.
3. Buée-Scherrer V, Condamines O, Mourton-Gilles C et al. (1996) AD2, a phosphorylation-dependent monoclonal antibody directed against tau proteins found in Alzheimer's disease. Brain Res Mol Brain Res 39: 79-88.
4. Sergeant N, Wattez A, Delacourte A (1999) Neurofibrillary degeneration in progressive supranuclear palsy and corticobasal degeneration: tau pathologies with exclusively "exon 10" isoforms. J Neurochem 72: 1243-1249.
5. Mulot SF, Hughes K, Woodgett JR et al. (1994) PHF-tau from Alzheimer's brain comprises four species on SDS-PAGE which can be mimicked by in vitro phosphorylation of human brain tau by glycogen synthase kinase-3 beta. FEBS Lett 349: 359-364.
6. Flament S, Delacourte A, Verny M et al. (1991) Abnormal Tau proteins in progressive supranuclear palsy. Similarities and differences with the neurofibrillary degeneration of the Alzheimer type. Acta Neuropathol 81: 591-596.
7. Delacourte A, Sergeant N, Wattez A et al. (1998) Vulnerable neuronal subsets in Alzheimer's and Pick's disease are distinguished by their tau isoform distribution and phosphorylation. Ann Neurol 43: 193-204.
8. Vermersch P, Sergeant N, Ruchoux MM et al. (1996) Specific tau variants in the brains of patients with myotonic dystrophy. Neurology 47: 711-717.
9. Mailliot C, Sergeant N, Bussière T et al. (1998) Phosphorylation of specific sets of tau isoforms reflects different neurofibrillary degeneration processes. FEBS Lett 433: 201-204.
10. Conrad C, Andreadis A, Trojanowski JQ et al. (1997) Genetic evidence for the involvement of tau in progressive supranuclear palsy. Ann Neurol 41: 277-281.
11. Evans DB, Rank KB, Bhattacharya K et al. (2000) Tau phosphorylation at Serine-396 and Serine-404 by human recombinant tau protein kinase II inhibits tau's ability to promote microtubule assembly. J Neurosci 20: 3937-3946.
12. Matsuo ES, Shin RW, Billingsley ML et al. (1994) Biopsy-derived adult human brain tau is phosphorylated at many of the same sites as Alzheimer's disease paired helical filament tau. Neuron 13: 989-1002.
13. Sergeant N, Bussière T, Vermersch P et al. (1995) Isoelectric point differentiates PHF-tau from biopsy-derived human brain tau proteins. Neuroreport 6:2217-2220.
14. Bussière T, Hof PR, Mailliot C et al. (1999) Phosphorylated serine422 on tau proteins is a pathological epitope found in several diseases with neurofibrillary degeneration. Acta Neuropathol 97: 221-230.
15. Probst A, Tolnay M, Langui D et al. (1996) Pick's disease: hyperphosphorylated tau protein segregates to the somatoaxonal compartment. Acta Neuropathol 92:588-596.
16. Cairns NJ, Atkinson PF, Hanger DP et al. (1997) Tau protein in the glial cytoplasmic inclusions of multiple system atrophy can be distinguished from abnormal tau in Alzheimer's disease. Neurosci Lett 230: 49-52.
17. Delacourte A, David JP, Sergeant N et al. (1999) The biochemical pathway of neurofibrillary degeneration in aging and Alzheimer's disease. Neurology 52: 1158-1165.

18. Working Group (1997) Consensus recommendations for the postmortem diagnosis of Alzheimer's disease. The National Institute on Aging, and Reagan Institute Working Group on Diagnostic Criteria for the Neuropathological Assessment of Alzheimer's Disease. Neurobiol Aging 18: S1-2.
19. Vermersch P, Robitaille Y, Bernier L et al. (1994) Biochemical mapping of neurofibrillary degeneration in a case of progressive supranuclear palsy: evidence for general cortical involvement. Acta Neuropathol 87: 572-577.
20. Buée Scherrer V, Hof PR, Buée L et al. (1996) Hyperphosphorylated tau proteins differentiate corticobasal degeneration and Pick's disease. Acta Neuropathol 91:351-359.
21. Delacourte A (2000) The biochemical pathway of neurofibrillary degeneration in aging and Alzheimer's disease. Neurology 54: 538.
22. Sahara N, Tomiyama T, Mori H (2000) Missense point mutations of tau to segregate with FTDP-17 exhibit site-specific effects on microtubule structure in COS cells: A novel action of R406W mutation. J Neurosci Res 60: 380-387.
23. Conrad C, Amano N, Andreadis A et al. (1998) Differences in a dinucleotide repeat polymorphism in the tau gene between Caucasian and Japanese populations: implication for progressive supranuclear palsy. Neurosci Lett 250: 135-137.

TAU GENE MUTATIONS AND TAU PATHOLOGY IN FRONTOTEMPORAL DEMENTIA AND PARKINSONISM LINKED TO CHROMOSOME 17

[1]Maria G. Spillantini and [2]Michel Goedert
[1]Brain Repair Centre and Department of Neurology, University of Cambridge, Forvie site, Robinson way, Cambridge CB2 2PY UK, MRC Laboratory of Molecular Biology, Hills Road, Cambridge CB2 2QH, UK.

1. INTRODUCTION

Frontotemporal dementia was first described by Arnold Pick in 1892[1]. In 1911, Alois Alzheimer described the neuropathological lesions characteristic of Pick's disease[2]. In the 1960's, these so-called Pick bodies were shown to contain abnormal filaments[3], which are now known to be made of hyperphosphorylated microtubule-associated protein tau[4,5]. They resemble the neurofibrillary lesions described by Alzheimer in 1907 in the disease subsequently named after him[6,7]. Filamentous inclusions made of tau protein are also a defining characteristic of neurodegenerative diseases, such as progressive supranuclear palsy (PSP) and corticobasal degeneration (CBD).

Frontotemporal dementias occur in familial forms and, more commonly, as sporadic diseases. Neuropathologically, they are characterized by a remarkably circumscribed atrophy of the frontal and temporal lobes of the cerebral cortex, often with additional, subcortical changes. In 1994, an autosomal-dominantly inherited familial form of frontotemporal dementia with parkinsonism was linked to chromosome 17q21.2[8]. This was followed by the identification of other familial forms of frontotemporal dementia that are linked to this region, resulting in the denomination frontotemporal dementia and parkinsonism linked to chromosome 17 (FTDP-17) for this class of disease[9]. A filamentous pathology made of hyperphosphorylated tau protein is a major neuropathological characteristic of FTDP-17[10]. Tau is a microtubule-associated protein that binds to microtubules and promotes microtubule assembly[7]. In diseases with tau inclusions, a fraction of tau does not bind to microtubules, and aggregates forming filaments. Tau filaments are space-occupying lesions that may interfere with a host of cellular processes, leading to the degeneration of affected nerve cells and glial cells. Importantly, the tau gene maps to the FTDP-17 locus on chromosome 17. Genetic linkage and neuropathology thus made *tau* a strong candidate gene for FTDP-17. The discovery of coding region and intronic mutations in *tau* has shown that the FTDP-17 locus is indeed the tau gene[11-13].

2. TAU GENE MUTATIONS IN FTDP-17

In adult human brain, six tau isoforms are produced from a single gene by alternative mRNA splicing (Figure 1a, A-F)[14-18]. They differ from each other by the presence or absence of 29- or 58-amino acid inserts located in the amino-terminal half and a 31 amino acid repeat located in the carboxy-terminal half. Inclusion of the latter, which is encoded by exon 10 of the tau gene, gives rise to the three tau isoforms (Figure 1a, D-F) with four repeats each. The other three isoforms (Figure 1a, A-C) have three repeats each. In normal cerebral cortex, similar levels of three-repeat and four-repeat tau isoforms are present. The repeats and some adjoining sequences constitute the microtubule-binding domains of tau[19,20].

Tau mutations in FTDP-17 are either missense, deletion or silent mutations in the coding region, or intronic mutations located close to the splice-donor site of the intron following the alternatively spliced exon 10 (Figure 1)[11-13,21-42,43,44]. Missense mutations are located in the microtubule-binding repeat region or close to it. Mutations in exon 9 (K257T and G272V), exon 12 (V337M) and exon 13 (G389R and R406W) affect all six tau isoforms (Figure 1a). By contrast, mutations in exon 10 (N279K, ΔK280, L284L, N296N, P301L, P301S, S305N and S305S) only affect tau isoforms with four microtubule-binding repeats or their levels of expression (Figure 1a). Most missense mutations reduce the ability of tau protein to interact with microtubules, as reflected by a reduction in the ability of mutant tau to promote microtubule assembly[23,30,38,45-49]. Mutations in exon 10 (ΔK280, P301L and P301S) produce the largest effects, with intermediate reductions for mutations in exon 9 (K257T and G272V) and 12 (V337M), and a smaller reduction for the G389R and R406W mutations in exon 13. Moreover, a number of missense mutations have a direct stimulatory effect on heparin-reduced assembly of tau into filaments[43,48,49]. This effect is particularly marked for the K257T, P301L and P301S mutations, with smaller effects for the G272V and V337M mutations. A study using synthetic peptides derived from each of the four microtubule-binding repeats of tau has also shown increased heparin-induced filament formation for the P301L mutation in the second repeat[49].

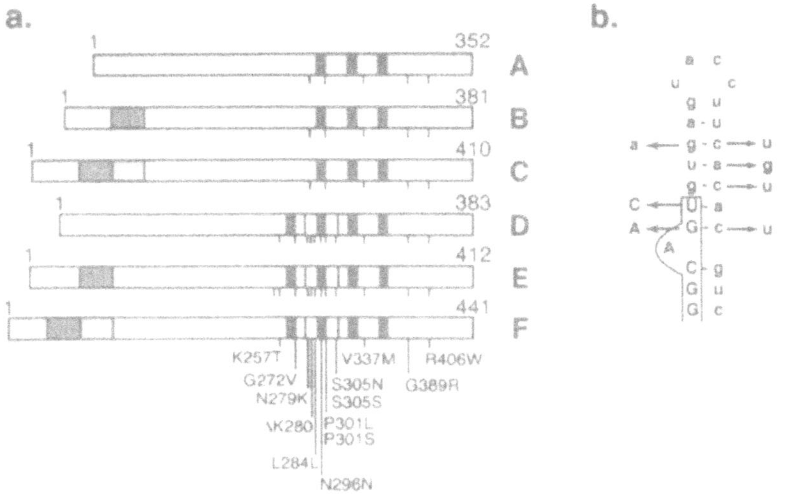

Figure 1. Mutations in the tau gene in frontotemporal dementia and parkinsonism linked to chromosome 17 (FTDP-17). **a.** Schematic diagram of the six tau isoforms (A-F) that are expressed in adult human brain. Alternatively spliced exons 2, 3 and 10 are shown in white dark gray, ligth gray and white boxes respectively. Black bars indicate the microtubule-binding repeats. Nine missense mutations, one deletion mutation and three silent mutations in the coding region are shown. Amino acid numbering corresponds to the 441 amino acid isoform of human brain tau. **b.** Stem-loop structure in the pre-mRNA at the boundary between exon 10 and the intron following exon 10. Seven mutations that reduce the stability of the stem-loop structure are shown. Exon sequences are shown in capital and intron sequences in lower-case letters.

Intronic mutations are located at positions +3, +12, +13, +14 and +16 of the intron following exon 10, with the first nucleotide of the splice-donor site taken as +1 (Figure 1b)[12,13,25,26,34,39,40]. Secondary structure predictions have suggested the presence of an RNA stem-loop structure at the exon 10-intron boundary that is disrupted by the intronic mutations[12,13]. In addition, the +3 mutation is predicted to lead to increased binding of U1 snRNA to the 5' splice site[13,14]. Exon trapping experiments have shown that intronic mutations increase splicing in of exon 10[12,26,29,32,39,50,51-53]. Increased production of transcripts encoding exon 10 has also been demonstrated in brain tissue from patients with tau intronic mutations[12,39]. This increase is in turn reflected in a change in the ratio of three-repeat to four-repeat tau isoforms, resulting in a net overproduction of four-repeat isoforms[13,26,34,39,44].

The proposed existence of a stem-loop structure at the boundary between exon 10 and the intron following exon 10 has received support from the determination of the three-dimensional structure of a 25 nucleotide-long RNA (extending from positions –5 to +19) by NMR spectroscopy (Figure 1b)[50-54]. It has shown that this sequence forms a stable, folded structure. The stem of this tau exon 10 splicing regulatory element consists of a single G-C base pair which is separated from a double helix of six base pairs by an unpaired adenine. The apical loop consists of six nucleotides that adopt multiple conformations in rapid exchange. The structure differs in several respects from the two proposed representations of the stem-loop[12,13]. Known intronic mutations are located in the upper part of the stem of the tau exon 10 splicing regulatory element (Figure 1b). All five mutations reduce the thermodynamic stability of the stem-loop structure, but to various extents [39,50]. The largest drop in melting temperature was observed for the +3 mutation. The +12 and +14 mutations also produced a large reduction in melting temperature, whereas the effects of the +13 and +16 mutations were smaller. The aminoglycoside antibiotic neomycin binds to the tau exon 10 splicing regulatory element and markedly increases the thermodynamic stability of both wild-type and mutant elements[54]. The differential reductions in stem-loop stability resulting from the various intronic mutations were reflected in the magnitude of increased splicing in of exon 10, as revealed by exon trapping[12,29,39,51-55].

The emerging picture is one of missense mutations that lead to a reduced ability of tau to interact with microtubules and to a stimulatory effect on filament assembly, and of intronic mutations whose primary effects are at the RNA level, resulting in an overproduction of tau isoforms with four microtubule-binding repeats. However, two missense mutations in exon 10 (N279K and S305N) deviate from this rule in that they do not lead to a reduction in the ability of tau to promote microtubule assembly[55]. Instead, they increase splicing in of exon 10, as is the case of the intronic mutations. Mutation N279K (AAT to AAG) creates a purine-rich splice enhancer sequence which explains its effects on exon trapping and soluble four-repeat tau in brain[22]. Similar findings have been obtained with the silent mutations L284L (CTT to CTC) and N296N (AAT to AAC) in exon 10 which are believed to disrupt an exon 10 splicing silencer sequence[29,44,56]. This work has uncovered the presence of sequence elements within exon 10 that regulate its alternative splicing. The S305N (AGT to AAT) mutation changes the last amino acid in exon 10[24]. This sequence forms part of the stem-loop structure, where the mutation produces a G to A transition at position –1 (Figure 1b). It is therefore not surprising that the S305N mutation leads to a reduction in the thermodynamic stability of the stem-loop structure and to a marked increase in the splicing in of exon 10[52,53,57]. Like the +3 mutation, the −1 mutation is also expected to lead to increased binding of U1 snRNA to the 5' splice site. The silent mutation S305S (AGT to AGC) produces a T to C transition at position 0 that disrupts the stem-loop structure, without a predicted effect on U1 snRNA binding (Figure 1b)[40]. As expected, it produced increased splicing in of exon 10 by exon trapping. Besides mutations in the intron following exon 10, additional pathogenic mutations may exist in other introns of the tau gene. Thus, a G to A transition at position +33 of the intron following exon 9 has been described in a patient with familial frontotemporal dementia[23]. It disrupts one of several (A/T) GGG repeats that may play a role in the regulation of the alternative splicing of exon 10.

3. NEUROPATHOLOGY OF FTDP-17 FAMILIES

All cases with tau gene mutations that have been examined to date have shown the presence of an abundant filamentous pathology made of hyperphosphorylated tau protein. Strikingly, the morphologies of tau filaments and their isoform compositions appear to be determined by whether tau mutations affect mRNA splicing of exon 10, or whether they are missense mutations located inside or outside exon 10[58] (Figure 2).

Mutations that affect splicing in of exon 10 lead to the formation of wide twisted ribbon-like filaments that only contain four-repeat tau isoforms. This has been shown in familial multiple system tauopathy with presenile dementia (MSTD) and a second family with the +3 intronic mutation[13,42,57], in FTD-Kumamoto with the +12 intronic mutation[39], as well as in familial progressive subcortical gliosis and Duke family 1684, both with the +16 intronic mutation[26,34]. Similar results have been obtained in pallido-ponto-nigral degeneration with the N279K mutation in exon 10 whose primary effects is at the RNA level[22,32,35]. The same may be true of families with the L284L, N296N, S305N and S305S mutations in exon 10 whose primary effect are also at the RNA level. In all these families, the tau pathology is widespread and present in both nerve cells and glial cells, with an abundant glial component (Figure 2b).

Mutations in exon 10 that do not affect alternative mRNA splicing lead to the formation of narrow twisted ribbons that contain four-repeat tau isoforms, with a small amount of the most abundant three-repeat isoform. This has been shown in Dutch family 1 and two other Dutch families and in an American family, all with the P301L mutation[27,36,59]. Based on electron microscopy of tissue sections, the same also appears to be true of the family with the P301S mutation[45]. At present, no neuropathological information is available for the family with thc ΔK280 mutation in exon 10. The P301L, P301S and ΔK280 mutations all lead to a markedly reduced ability of four-repeat tau to promote microtubule assembly[41]. By exon trapping, the P301L and P301S mutations have no effect on the splicing in of exon 10. By contrast, the ΔK280 mutation leads to reduced splicing in of exon 10, suggesting that its primary effect might be the overproduction of three-repeat tau, and not the reduced ability of four-repeat tau to interact with mirotubules[29]. Clarification of this issue must await the availability of frozen brain tissue from an individual with the ΔK280 mutation. In brain tissue from individuals with the P301L and P301S mutations, tau pathology is widespread and present in both nerve cells and glial cells. When compared with mutations that affect the splicing in of exon 10, the glial component appears to be less pronounced.

Coding region mutations located outside exon 10 lead to a tau pathology that is neuronal, without a significant glial component (Figure 2a). Some of these mutations lead to the formation of paired helical and straight filaments that contain all six tau isoforms, like the tau filaments of Alzheimer's disease [60]. This has been shown for Seattle family A with the V337M mutation in exon 12 and for a family with the R406W mutation in exon 13[36,61]. In both cases, the morphologies of tau filaments have been found to be indistinguishable from those of Alzheimer's disease (Figure 2c). By contrast, the G389R mutation in exon 13 produces tau filament morphologies and a pattern of tau bands that resemble the characteristics of Pick's disease[38,62,63]. Based on light microscopic staining, the G272V mutation in exon 9 leads to the formation of numerous Pick body-like inclusions[59]. Similarly, the K257T mutation in exon 9, leads to the formation of Pick's bodies with slightly twisted filaments and a pattern of pathological tau bands similar to sporadic Pick's disease[43] (Figure 2d). These findings indicate that depending on the positions of tau missense mutations in exons 9, 12 and 13, and perhaps the nature of these mutations, a filamentous tau pathology ensues which resembles that of either Alzheimer's disease or Pick's disease.

4. PATHOGENESIS OF FTDP-17

The pathway leading from a mutation in the tau gene to neurodegeneration is unknown. The likely primary effect of most missense mutations is a reduced ability of mutant tau to interact with microtubules. It may be equivalent to a partial loss of function with resultant microtubule destabilization and deleterious effects on cellular processes, such as rapid axonal transport. However, in the case of the intronic mutations and the

N279K, L284L, N296N, S305N and S305S mutations in exon 10, this appears unlikely. The net effect of these mutations is increased splicing in of exon 10, leading to a change in the ratio of three-repeat to four-repeat tau isoforms, and resulting in an overproduction of four-repeat tau. It is well known that four-repeat tau possesses a greater ability to interact with microtubules than three-repeat tau[17]. It is therefore possible that in cases of FTDP-17 with intronic mutations and coding region mutations that act at the RNA level, microtubules are more stable than in brain from control individuals. Moreover, missense mutations in exon 10 will only affect 20-25% of tau molecules, with 75-80% of tau being normal.

It is possible, however, that a correct ratio of wild-type three-repeat to four-repeat tau is essential for the normal function of tau in human brain. An alternative hypothesis is that a partial loss of function of tau is necessary for setting in motion the mechanisms that will ultimately lead to filament assembly. Earlier work has suggested that three-repeat and four-repeat tau isoforms may bind to different sites on microtubules[20]. Overproduction of tau isoforms with four repeats may result in an excess of tau over available binding sites on microtubules, thus creating a gain of toxic function similar to that of most missense mutations, with unbound excess tau available for assembly into filaments.

Where studied, pathological tau from FTDP-17 brain is hyperphosphorylated[7,10]. Since known mutations in tau do not create additional phosphorylation sites (with the possible exception of the P301S and K257T mutations), hyperphosphorylation of tau is probably an event downstream of the primary effects of the mutation and may be a consequence of the partial loss of function. However, some missense mutations may indirectly affect the phosphorylation state of tau. Thus, in transfected cells, tau protein with the R406W mutation displays only little phosphorylation at T231, S396 and S404, in contrast to wild-type tau and tau with the P301L and V337M mutations[46,64]. Hyperphosphorylation of tau probably reinforces the primary effects of the mutations, since it is well established that hyperphosphorylated tau is unable to bind to microtubules and to promote microtubule assembly[65,66]. At present, there is no experimental evidence linking hyperphosphorylation of tau to filament assembly, and it is unclear whether hyperphosphorylation is either necessary or sufficient for assembly.

Sulphated glycosaminoglycans and RNA induce the bulk assembly of non-phosphorylated, recombinant tau protein into Alzheimer-like filaments *in vitro*[67-70]. This work has produced robust methods for the assembly of full-length tau into filaments. However, the mechanisms that lead to the assembly of tau into filaments in brain remain to be discovered. It is possible that a reduced ability of tau to interact with microtubules, which could have several different causes, is a necessary step for filament formation. Assembly is an energetically unfavourable, nucleation-dependent process that requires a critical concentration of tau[67,71]. Many cells may have levels of tau below the critical concentration. Other cells may have effective mechanisms for preventing the formation of tau nuclei, or may be able to degrade them once they have formed.

Insufficient protective mechanisms and high tau concentrations may underlie the selective degeneration of nerve cells and glial cells, which is especially striking in FTDP-17, with the characteristic, sometimes unilateral, razor-sharp demarcations between affected and unaffected areas in cerebral cortex. Similar factors may also underpin the late ages of onset of this and other diseases with filamentous tau protein deposits. Protective factors, such as proteases that degrade nucleation products, may be effective throughout much of life. However, as nerve cells age, these mechanisms may become less effective and the balance may tilt in favour of filament formation.

The significance of the different filament morphologies observed in FTDP-17 is not clear. Is is known that the repeat region of tau forms the densely packed core of paired helical and straight filaments of Alzheimer's disease, with the amino- and carboxy-terminal parts of the molecule forming a protease-sensitive coat[60,72]. Also, for filaments assembled in vitro in the presence of sulphated glycosaminoglycans, the morphology of the filaments depends on the number of repeats in the tau isoform used[67]. Thus, mutations in the repeat region or a change in the relative amounts of three- and four-repeat isoforms could well influence filament morphology. However, treatment of paired helical filaments, which contain all six tau isoforms, with acid leads to untwisted, ribbon-like filaments like those seen in cases of FTDP-17 with mutations in the intron following exon 10, suggesting a close similarity in packing of tau molecules in the various structures[73]. The most important aspect may be the extended filamentous nature of the assemblies that can make them resistant to proteolysis and the deleterious effects that this has on intracellular processes, rather than the detailed morphology of the different filaments.

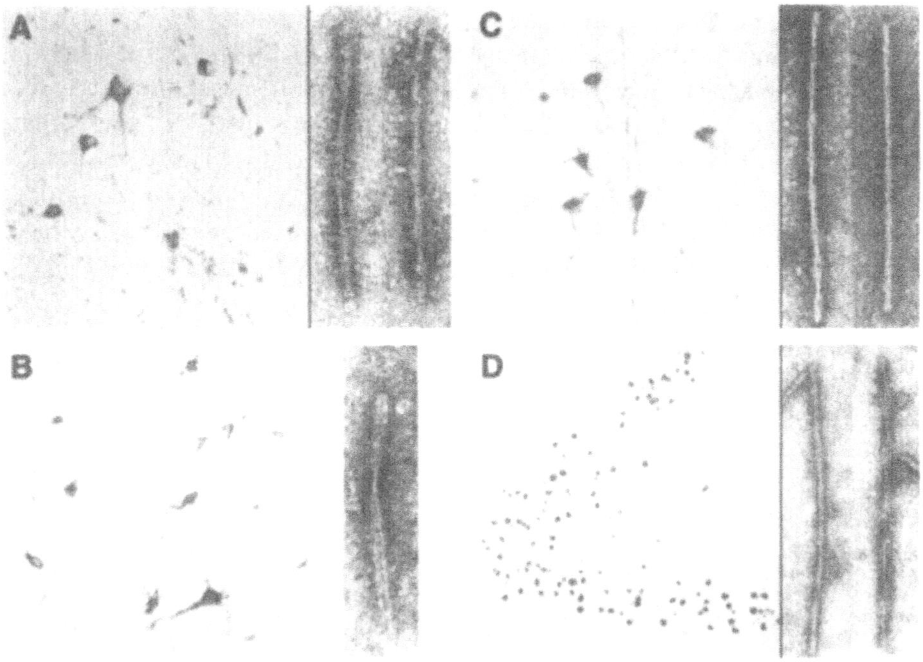

Figure 2. FTDP-17 tau pathologies. **A-D** Tau pathologies of FTDP-17, as revealed by immunohistochemistry for hyperhosphorylated tau protein and the morphologies of isolated tau filaments. **A** The P301L mutation in exon 10 give rise to a neuronal and glial tau pathology. Tau filaments consist of narrow twisted ribbons (left) as the majority species and rope-like filaments (right) as the minority species. Biochemically and structurally the tau pathology in this case is similar to sporadic progressive supranuclear palsy. **B** Mutations in the intron following exon 10 give rise to a neuronal and glial tau pathology. Tau filaments consist of wide twisted ribbons made of four-repeat tau isoforms. The glial pathology is more extensive than A. The tau pathology is similar to corticobasal degeneration. **C** The V337M mutation in exon 12 give rise to a neuronal tau pathology. Tau filaments consist of paired helical (left) and straight (right) filaments, like the filaments of Alzheimer's disease. They contain all six human brain tau isoforms. Paired helical filaments constitute the majority species. The R406W mutation in exon 13 gives rise to a similar tau pathology. The pathology in cases with these two mutations is very similar to that of Alzheimer's disease. **D** The K257T mutation in exon 9 gives rise to a neuronal tau pathology. Tau filaments consist of straight filaments (left) as the majority species and twisted filaments (right) as the majority species. The tau pathology resembles that of Pick's disease. Filaments consist of three-repeat and four-repeat tau isoforms. The G389R mutation in exon 13 gives a similar pathology.

5. CONCLUSIONS

The discovery of mutations in the tau gene has established that dysfunction of tau protein causes neurodegeneration. Most unexpectedly, it has shown that similar levels of three-repeat and four-repeat tau isoforms are essential for preventing neurodegeneration and dementia. The finding that overproduction of four-repeat tau leads to its assembly into twisted ribbons and causes disease may shed light on the pathogenesis of progressive supranuclear palsy (PSP) and corticobasal degeneration (CBD), two largely sporadic neurodegenerative diseases. Neuropathologically, they are characterised by an abundant filamentous pathology that is comprised predominantly of four-repeat tau isoforms[74-76]. A recent study has reported an increase in exon 10-containing transcripts in brainstem in PSP[77]. It remains to be seen whether this is reflected by increased levels of soluble four-repeat tau isoforms. An association between PSP and homozygosity of a common allele at a dinucleotide repeat in the intron following exon 9 of the tau gene has been described[78]. More recently, two common tau haplotypes that differ at the nucleotide level, but not at the level of the protein coding sequence, have been reported[79]. Homozygosity of the more common allele H1 appears to predispose to PSP. Taken together, this work suggests that PSP may be caused by an imbalance between three-repeat and four-repeat tau isoforms, analogous to the FTDP-17 cases with mutations in the intron following *tau* exon 10. The same appears to be true of CBD[7]. It is noteworthy that a patient with the S305S mutation in *tau* presented with the symptoms of PSP[40], whereas a patient with the P301S mutation in *tau* showed the clinical signs of CBD[30]. Moreover, the N296N mutation in tau leads to a clinical and neuropathological picture that resembles CBD[44].

The finding that some mutations in *tau* exons 9 and 13 give rise to a clinical picture and a neuropathology that are similar to those of Pick's disease indicates their relevance for an understanding of the aetiology and pathogenesis of Pick's disease. These mutations affect all six tau isoforms and where studied, the tau filaments contain three-repeat and four-repeat tau isoforms[38,43]. This is in apparent contrast to sporadic Pick's disease, which is believed to be characterised by the assembly of three-repeat, but not four-repeat, tau into filaments[80,81].

All cases of FTDP-17 examined to date are characterised by an abundant filamentous tau pathology. The V337M and R406W mutations lead to a tau pathology that is identical to that of Alzheimer's disease in its biochemical and morphological characteristics[36,61]. It follows that in Alzheimer's disease nerve cells also degenerate as the direct result of the development of a filamentous tau pathology. Unlike in Alzheimer's disease, abundant neuritic plaques are not a general characteristic of FTDP-17. However, isolated cases of FTDP-17 with abundant amyloid deposits have been described[29,36]. Although one cannot exclude concomitant Alzheimer's disease pathology in these patients, future work may uncover mechanisms by which the development of tau pathology can lead to deposition of Abeta amyloid.

Prevention or reduction of tau pathology is likely to be therapeutically beneficial. Current evidence suggests that binding of tau to microtubules and self-assembly into filaments are mutually exclusive. Therefore, increasing the ability of abnormally hyperphosphorylated tau to interact with microtubules may be of benefit. This could in principle be achieved by using specific protein kinase inhibitors, proteins that bind to protein kinases and inactivate them or a *cis-trans* prolyl isomerase, such as Pin1[82-84]. The latter has been shown to bind to hyperphosphorylated tau protein *in vitro* and to restore its ability to interact with microtubules[84]. Naturally occurring osmolytes, such as trimethylamine N-oxide, have also been shown to restore the ability of phosphorylated tau to interact with microtubules[85]. Assembly of tau into filaments is a nucleation-dependent process that is strongly concentration dependent. Therefore a reduction of the cellular levels of tau protein must be a therapeutic target. This is especially true of tau mutations whose primary effect is at the RNA level. Small organic compounds that specifically stabilise the stem-loop structure at the boundary between tau exon 10 and the intron following exon 10 may be able to restore the disturbed ratio between three-repeat and four-repeat tau isoforms[54]. Compounds that prevent the assembly of tau protein into filaments can be identified using existing methods for *in vitro* filament formation. Validation of their usefulness will require the development of experimental animal models of tau filament formation. Transgenic mouse lines overexpressing individual wild-type three-repeat or four-repeat human tau isoforms within nerve cells have exhibited signs of neurodegeneration[86-91]. Mice from the highest expressing lines showed marked axonopathy and amyotrophy, accompanied by numerous neurofilament-positive spheroids. No clear evidence of tau filaments or nerve cell loss was presented. This work has shown that the accumulation of non-filamentous human tau protein is sufficient to cause some nerve cell dysfunction. However, the pathological changes observed were much less pronounced than those found in FTDP-17.

The discovery of mutations in the tau gene in FTDP-17 is leading to the production of transgenic mouse lines expressing mutant human tau protein in nerve cells and glial cells. A first study reporting expression of the shortest four-repeat tau isoform with the P301L mutation has shown the formation of filaments made of hyperphosphorylated tau protein and the degeneration of affected nerve cells[92]. Neurofibrillary tangles, axonopathy, amyotrophy, spheroids and reactive gliosis were also present. These findings are consistent with the view that formation of tau filaments is a major determinant of nerve cell degeneration in the tauopathies.

REFERENCES

1. Pick A (1892) Über die Beziehungen der senilen Hirnatrophie zur Aphasie. Prager Med Wochenschr 16: 765-767.
2. Alzheimer A (1911) Über eigenartige Krankheitsfälle des Späteren Alters. Z ges Neurol Psychiat 4: 356-385.
3. Rewcastle NB, Ball MJ (1968) Electron microscopic structure of the inclusion bodies in Pick's disease. Neurology 18: 1205-1213.
4. Pollock NJ, Mirra SS, Binder LI et al. (1986) Filamentous aggregates in Pick's disease, progressive supranuclear palsy, and Alzheimer's disease share antigenic determinants with microtubule-associated protein tau. Lancet II: 1211.
5. Probst A, Tolnay M, Langui D et al. (1996) Pick's disease: hyperphosphorylated tau segregates to the somatoaxonal compartment. Acta Neuropathol 92: 588-596.
6. Alzheimer A (1907) Über eine eigenartige Erkrankung der Hirnrinde. Allg Z Psychiat Psych Gerichtl Med 64: 146-148.
7. Spillantini MG, Goedert M (1998) Tau protein pathology in neurodegenerative diseases Trends Neurosci 21: 428-433.
8. Wilhelmsen KC, Lynch T, Pavlou E et al. (1994) Localization of disinhibition-dementia-parkinsonism-amyotrophy complex to 17q21-22. Am J Hum Genet 55: 1159-1165.
9. Foster NL, Wilhelmsen KC, Sima AAF et al. (1997) Frontotemporal dementia and parkinsonism linked to chromosome 17: a consensus statement. Ann Neurol 41: 706-715.
10. Spillantini MG, Bird TD, Ghetti B (1998) Frontotemporal dementia and Parkinsonism linked to chromosome 17: a new group of tauopathies. Brain Pathol 8: 387-402.
11. Poorkaj P, Bird TD, Wijsman E et al. (1998) Tau is a candidate gene for chromosome 17 frontotemporal dementia. Ann Neurol 43: 815-825.
12. Hutton M, Lendon CL, Rizzu P et al. (1998) Association of missense and 5'-splice-site mutations in tau with the inherited dementia FTDP-17. Nature 393: 702-705.
13. Spillantini MG, Murrell JR, Goedert M et al. (1998) Mutation in the tau gene in familial multiple system tauopathy with presenile dementia. Proc Natl Acad Sci USA 95: 7737-7741.
14. Goedert M, Wischik CM, Crowther RA et al. (1988) Cloning and sequencing of the cDNA encoding a core protein of the paired helical filament of Alzheimer's disease. Proc Natl Acad Sci USA 85: 4051-4055.
15. Goedert M, Spillantini MG, Potier MC et al. (1989) Cloning and sequencing of the cDNA encoding an isoform of microtubule associated protein tau containing four tandem repeats; differential expression of tau protein mRNAs in human brain. EMBO J 8: 393-399.
16. Goedert M, Spillantini MG, Jakes R et al. (1989) Multiple isoforms of human microtubule-associated protein tau: sequences and localization in neurofibrillary tangles of Alzheimer's disease. Neuron 3: 519-526.
17. Goedert M, Jakes R (1990) Expression of separate isoforms of human tau protein: correlation with the tau pattern in brain and effects on tubulin polymerization. EMBO J 9: 4225-4230.
18. Andreadis A, Brown MW, Kosik KS (1992) Structure and novel exons of the human tau gene. Biochemistry 31: 10626-10633.
19. Gustke N, Trinczek B, Biernat J et al. (1994) Domains of tau protein and interaction with microtubules. Biochemistry 33: 9511-9522.

20. Goode BL, Feinstein SC (1994) Identification of a novel microtubule-binding and assembly domain in the developmentally regulated inter-repeat region of tau. J Cell Biol 124: 769-782.
21. Dumanchin C, Camuzat A, Campion D et al. (1998) Segregation of a missense mutation in the microtubule-associated protein tau gene with familial frontotemporal dementia and parkinsonism. Hum Mol Genet 7: 1825-1829.
22. Clark LN, Poorkaj P, Wszolek Z et al. (1998) Pathogenic implications of mutations in the tau gene in pallido-ponto-nigral degeneration and related neurodegenerative disorders linked to chromosome 17. Proc Natl Acad Sci USA 95: 13103-13107.
23. Rizzu P, Van Swieten JC, Joosse M et al. (1999) High prevalence of mutations in the microtubule associated protein tau in a population study of frontotemporal dementia in the Netherlands. Am J Hum Genet 64: 414-421.
24. Iijima M, Tabira T, Poorkaj P et al. (1999) A distinct familial presenile dementia with a novel missense mutation in the tau gene. Neuroreport 10: 497-501.
25. Morris HR, Perez-Tur J, Janssen JC et al. (1999) Mutations in the tau exon 10 splice site region in familial frontotemporal dementia. Ann Neurol 45: 270-271.
26. Goedert M, Spillantini MG, Crowther RA et al. (1999) Tau gene mutation in familial progressive subcortical gliosis. Nat Med 5: 454-457.
27. Mirra SS, Murrell JR, Gearing M et al. (1999) Tau pathology in a family with dementia and a P301L mutation in tau. J Neuropathol Exp Neurol 58: 333-345.
28. Bird TD, Nochlin D, Poorkaj P et al. (1999) A clinical pathological comparison of three families with frontotemporal dementia and identical mutations in the tau gene (P301L). Brain 122: 741-756.
29. D'Souza I, Poorkaj P, Hong M et al. (1999) Missense and silent tau gene mutations cause frontotemporal dementia and parkinsonism - chromosome 17, by affecting multiple alternative RNA splicing regulatory elements. Proc Natl Acad Sci USA 96: 5598-5603.
30. Bugiani O, Murrell JR, Giaccone G et al. (1999) Frontotemporal dementia and corticobasal degeneration in a family with a P301S mutation in Tau. J Neuropathol Exp Neurol 58: 667-677.
31. Nasreddine ZS, Loginov M, Clark LN et al (1999) From genotype to phenotype: a clinical, pathological and biochemical investigation of frontotemporal dementia and parkinsonism (FTDP-17) caused by the P301L tau mutation. Ann Neurol 45: 704-715.
32. Delisle MB, Murrell J R, Richardson R et al. (1999) A mutation at codon 279 (N279K) in exon 10 of the Tau gene causes a tauopathy with dementia and supranuclear palsy. Acta Neuropathol 98: 62-77.
33. Houlden H, Baker M, Adamson J et al. (1999) Frequency of tau mutations in three series of non-Alzheimer's degenerative dementia. Ann Neurol 46: 243-248.
34. Hulette CM, Pericak-Vance MA, Roses AD et al. (1999) Neuropathological features of frontotemporal dementia and parkinsonism linked to chromosome 17q21-22 (FTDP-17): Duke family 1684. J Neuropathol Exp Neurol 58: 859-866.
35. Yasuda M, Kawamata T, Komure O et al. (1999) A mutation in the microtubule-associated protein tau in pallido-nigro-luysian degeneration. Neurology 53: 864-868.
36. Van Swieten JC, Stevens M, Rosso SM et al. (1999) Phenotypic variation in hereditary frontotemporal dementia with tau mutations. Ann Neurol 46: 617-626.
37. Sperfeld AD, Collatz MB, Baier H et al. (1999) FTDP-17: an early-onset phenotype with parkinsonism and epileptic seizures caused by a novel mutation. Ann Neurol 46: 708-715.

38. Murrell JR, Spillantini MG, Zolo P et al. (1999) Tau gene mutation G389R causes a tauopathy with abundant Pick body-like inclusions and axonal deposits. J Neuropathol Exp Neurol 58: 1207-1226.
39. Yasuda M, Takamatsu J, D'Souza I et al. (2000) A novel mutation at position +12 in the intron following exon 10 of the tau gene in familial frontotemporal dementia (FTD-Kumamoto). Ann Neurol 47: 422-429.
40. Stanford PM, Halliday GM, Brooks WS et al. (2000) Progressive supranuclear palsy pathology caused by a novel silent mutation in exon 10 of the tau gene. Expansion of the disease phenotype caused by tau gene mutations. Brain 123: 880-893.
41. Hong M, Zhukareva V, Vogelsberg-Ragaglia V et al. (1998) Mutation specific functional impairments in distinct tau isoforms of hereditary FTDP-17. Science 282: 1914-1917.
42. Tolnay M, Spillantini MG, Rizzini C et al. (2000) A new case of frontotemporal dementia and parkinsonism resulting from an intron 10 +3-splice site mutation in the tau gene: clinical and pathological features. Neuropathol Appl Neurobiol 26: 368-378.
43. Rizzini C, Goedert M, Crowther RA et al. (2000) Tau gene mutation K257T causes a tauopathy similar to Pick's disease. J Neuropathol Exp Neurol, in press.
44. Spillantini MG, Yoshida H, Rizzini C et al. (2000) A novel Tau mutation (N296N) in familial dementia with swollen achromatic neurons and corticobasal inclusion bodies. Ann Neurol, in press.
45. Hasegawa M, Smith MJ, Goedert M (1998) Tau proteins with FTDP-17 mutations have a reduced ability to promote microtubule assembly. FEBS Lett 437: 207-210.
46. Dayanandan R, Van Slegtenhorst M, Mack TGA et al. (1999) Mutations in tau reduce its microtubule binding properties in intact cells and affects its phosphorylation. FEBS Lett 446: 228-232.
47. DeTure M, Ko L-W, Yen S et al. (2000) Missense tau mutations identified in FTDP-17 have a small effect on tau microtubule interactions. Brain Res 853: 5-14.
48. Nacharaju P, Lewis J, Easson C et al. (1999) Accelerated filaments formation from tau protein with specific FTDP-17 mutations. FEBS Lett 447: 195-199.
49. Goedert M, Jakes R, Crowther RA (1999) Effects of frontotemporal dementia FTDP-17 mutations on heparin-induced assembly of tau filaments. FEBS Lett 450: 306-311.
50. Varani L, Hasegawa M, Spillantini MG et al. (1999) Structure of tau exon 10 splicing regulatory element RNA and destabilization by mutations of frontotemporal dementia and parkinsonism linked to chromosome 17. Proc Natl Acad Sci USA 96: 8229-8234.
51. Grover A, Houlden H, Baker M et al. (1999) 5' Splice site mutations in tau associated with the inherited dementia FTDP-17 affect a stem loop structure that regulates alternative splicing of exon 10. J Biol Chem 274: 15134-15143.
52. Gao Q-S, Memmott J, Lafyatis R et al. (2000) Complex regulation of tau exon 10, whose missplicing causes frontotemporal dementia. J Neurochem 74: 490-500.
53. Jiang Z, Cote J, Kwon JM et al. (2000) Aberrant splicing of tau pre-mRNA caused by intronic mutations associated with the inherited dementia frontotemporal dementia with parkinsonism linked to chromosome 17. Mol Cell Biol 20: 4036-4048.
54. Varani L, Spillantini MG, Goedert M et al. (2000) Structural basis for recognition of the RNA major groove in the tau exon 10 splicing regulatory element by aminoglycoside antibiotics. Nucl Ac Res 28: 710-719.
55. Hasegawa M, Smith MJ, Iijima M et al. (1999) FTDP-17 mutations N279K and S305N in tau produce increased splicing of exon 10. FEBS Lett 443: 93-96.

56. D'Souza I, Schellenberg GD (2000) Determinants of 4-repeat tau expression. Coordination between enhancing and inhibitory splicing sequences for exon 10 inclusion. J Biol Chem 275: 17700-17709.
57. Spillantini MG, Goedert M, Crowther RA et al. (1997) Familial multiple system tauopathy with presenile dementia: a disease with abundant neuronal and glial tau filaments. Proc Natl Acad Sci USA 94: 4113-4118.
58. Goedert M, Crowther RA, Spillantini MG (1998) Tau mutations cause frontotemporal dementias. Neuron 21: 955-958.
59. Spillantini MG, Crowther RA, Kamphorst W et al. (1998) Tau pathology in two Dutch families with tau mutations in the microtubule binding region of tau. Am J Pathol 153: 1359-1363.
60. Goedert M, Spillantini MG, Cairns NJ et al. (1992) Tau proteins of Alzheimer paired helical filaments: abnormal phosphorylation of all six brain isoforms. Neuron 8: 159-168.
61. Spillantini MG, Crowther RA, Goedert M (1996) Comparison of the neurofibrillary pathology in Alzheimer's disease and familial presenile dementia with tangles. Acta Neuropathol 92: 42-48.
62. Kato S, Nakamura H (1990) Presence of two different fibril subtypes in the Pick body: an immunoelectron microscopic study. Acta Neuropathol 81: 125-129.
63. Murayama S, Mori H, Ihara Y et al. (1990) Immunohistochemical and ultrastructural studies of Pick's disease. Ann Neurol 27: 394-404.
64. Matsumura N, Yamazaki T, Ihara Y (1999) Stable expression in Chinese hamster ovary cells of mutated tau genes causing frontotemporal dementia and parkinsonism linked to chromosome 17 (FTDP-17). Am J Pathol 154: 1649-1659.
65. Bramblett GT, Goedert M, Jakes R et al. (1993) Abnormal tau phosphorylation at Ser396 in Alzheimer's disease recapitulates development and contributes to reduced microtubule binding. Neuron 19: 1089-1099.
66. Yoshida H, Ihara Y (1993) Tau in paired helical filament is functionally distinct from fetal tau: assembly incompetence of paired helical filament tau. J. Neurochem 61: 1183-1186.
67. Goedert M, Jakes R, Spillantini MG et al. (1996) Assembly of microtubule-associated protein tau into Alzheimer-like filaments induced by sulphated glycosaminoglycans. Nature 383: 550-553.
68. Perez M, Valpuesta JM, Medina M et al. (1996) Polymerization of tau into filaments in the presence of heparin: the minimal sequence required for tau-tau interactions. J Neurochem 67: 1183-1190.
69. Kampers T, Friedhoff P, Biernat J et al. (1996) RNA stimulates aggregation of microtubule-associated protein tau into Alzheimer-like paired helical filaments. FEBS Lett 399: 344-349.
70. Hasegawa M, Crowther RA, Jakes R et al. (1997) Alzheimer-like changes in microtubule-associated protein tau induced by sulfated glycosaminoglycans. Inhibition of microtubule binding, stimulation of phosphorylation and filament assembly depend on the degree of sulfation. J Biol Chem 272: 33118-33124.
71. Friedhoff P, von Bergen M, Mandelkow EM et al. (1998) A nucleated assembly mechanism of Alzheimer paired helical filaments. Proc Natl Acad Sci USA 95: 15712-15717.
72. Wischik CM, Novak M, Thogersen HC et al. (1988) Isolation of a fragment of tau derived from the core of the paired helical filament of Alzheimer disease. Proc Natl Acad Sci USA 85: 4506-4510.

73. Crowther RA (1991) Structural aspects of pathology in Alzheimer's disease. Biochem Biophys Acta 1069: 1-9.
74. Flament S, Delacourte A, Vernay M et al. (1991) Abnormal tau proteins in progressive supranuclear palsy. Similarities and differences with the neurofibrillary degeneration of the Alzheimer type. Acta Neuropathol 81: 591-596.
75. Ksiezak-Reding H, Morgan K, Mattiace LA et al. (1994) Ultrastructure and biochemical composition of paired helical filaments in corticobasal degeneration. Am J Pathol 145: 1496-1508.
76. Sergeant N, Wattez A, Delacourte A (1999) Neurofibrillary degeneration in progressive supranuclear palsy and corticobasal degeneration: Tau pathologies with exclusively "exon 10" isoforms. J Neurochem 72: 1243-1249.
75. Chambers CB, Lee JM, Troncoso JC et al. (1999) Overexpression of four-repeat tau mRNA isoforms in progressive supranuclear palsy but not in Alzheimer's disease. Ann Neurol 46: 325-332.
76. Conrad C, Andreadis A, Trojanowski JQ et al. (1997) Genetic evidence for the involvement of tau in progressive supranuclear palsy. Ann Neurol 41: 277-281.
77. Baker M, Litvan I, Houlden H et al. (1999) Association of an extended haplotype in the tau gene with progressive supranuclear palsy. Hum Mol Genet 8: 711-715.
78. Delacourte A, Robitaille Y, Sergeant N et al. (1996) Specifc pathological tau protein variants characterize Pick's disease. J Neuropathol Exp Neurol 55: 159-168.
79. Delacourte A, Sergeant N, Wattez A et al. (1998) Vulnerable subsets in Alzheimer's and Pick's disease are distinguished by their tau isoform distribution and phosphorylation. Ann Neurol 43: 193-204.
80. Cohen P, Goedert M (1999) Engineering protein kinases with distinct nucleotide specificities and inhibitor sensitivities by mutation of a single amino acid. Chem Biol 5: R161-R164.
81. Thomas GM, Frame S, Goedert M et al. (1999) A GSK3-binding peptide from FRAT1 selectively inhibits GSK3-catalysed phosphorylation of axin and ß-catenin. FEBS Lett. 458: 247-251.
82. Lu P-J, Wulf G, Zhou XZ et al. (1999) The prolyl isomerase Pin1 restores the function of Alzheimer-associated phosphorylated tau protein. Nature 399: 784-788.
83. Tseng H-C, Lu Q, Henderson E, Graves DJ (1999) Phosphorylated tau can promote tubulin assembly. Proc Natl Acad Sci USA 96: 9503-9508.
84. Götz J, Probst A, Spillantini MG et al. (1995) Somatodendritic localization and hyperphosphorylation of tau protein in transgenic mice expressing the longest human brain tau isoform. EMBO J 14: 1304-1313.
85. Brion JP, Tremp G, Octave JN (1999) Transgenic expression of the shortest tau affects its compartmentalization and its phosphorylation as in the pretangle stage of Alzheimer's disease. Am J Pathol 154: 255-270.
86. Ishihara T, Hong M, Zhang B et al. (1999) Age-dependent emergence and progression of a tauopathy in transgenic mice overexpressing the shortest human tau isoform. Neuron 24: 751-762.
87. Spittaels K, Van den Haute C, Van Dorpe J et al. (1999) Prominent axonopathy in the brain and spinal cord of transgenic mice overexpressing four-repeat human tau protein. Am J Pathol 155: 2153-2165.
88. Probst A, Götz J, Wiederhold K-H et al. (2000) Axonopathy and amyotrophy in mice transgenic for human four-repeat tau protein. Acta Neuropathol 99: 469-481.

89. Duff K, Knoght H, Refolo LM et al. (2000) Characterization of pathology in transgenic mice over-expressing human genomic and cDNA tau transgenes. Neurobiol Dis 8: 87-98.
90. Lewis J, McGowan E, Rockwood J et al. (2000) Neurofibrillary tangles, amyotrophy and progressive motor disturbance in mice expressing mutant (P301L) tau protein. Nat Genet 25: 402-425.

ARGYROPHILIC GRAIN DISEASE
A frequent dementing disorder in aged patients

[1]Markus Tolnay, [2]Andreas U. Monsch and [2]Alphonse Probst
[1]Institute of Pathology, Division of Neuropathology, Basel University, Basel, Switzerland:
[2]Geriatric University Clinic, Basel University, Basel, Switzerland

1. INTRODUCTION

Prospective clinicopathological studies have confirmed that Alzheimer's disease (AD) is the most common neurodegenerative cause of late-onset dementia which affects 15-20 million subjects world-wide and constitutes the fourth leading cause of death in industrialised countries.

AD is characterized clinically by a progressive loss of memory and cognitive functions, resulting in a severe dementia. Neuropathologically AD is defined by the accumulation of two types of insoluble fibrous material: (i) extracellular amyloid (Aß) protein in the shape of senile plaques (SP) and (ii) intracellular neurofibrillary lesions (NFL) made of abnormally and hyperphosphorylated tau protein[1,2]. During the last 10 years there has been an increasing recognition of so called non-Alzheimer degenerative disorders accounting for between 10 - 20% of all cases of dementias. Among them there is a subset of disorders sharing filamentous tau protein deposits in neurons and/or glial cells as their main pathological characteristic, therefore termed tauopathies[3,4]. They are characterized clinically by dementia (with or without motor disturbances) and include among others Pick's disease (PiD), corticobasal degeneration (CBD), progressive supranuclear palsy (PSP) and familial frontotemporal dementia and parkinsonism linked to chromosome 17 (FTDP-17) (for recent reviews see[4,5,6]). Among the tauopathies, argyrophilic grain disease (AgD) still represents a less well known disorder although recent work clearly demonstrated its high prevalence in advanced age and its potential to cause severe brain dysfunction[7-9].

Tauopathies represent a diagnostic challenge for both clinicians and neuropathologists. A substantial overlap of pathological features may exist between the tauopathies with many cellular lesions being encountered in more than one disorder. However, some morphological features like Pick bodies of PiD[10], "astrocytic plaques" of CBD[11], tufted astrocytes of PSP[12] or argyrophilic grains (ArGs) of AgD[13,14] may be considered disease specific. Moreover, the deterioration of the brain caused by each tauopathy is neither diffuse nor random. Rather, specific subsets of nerve cells, usually belonging to more than one neuronal system and/or local subsets of glial cells, become systematically affected and often perish according to a predictable sequence, whereas adjoining cells remain unscathed or suffer only secondary damage. Accordingly, a typical pattern of nerve cell degeneration which usually remains consistent across cases develops for each of these disorders. Owing to these features, a given tauopathy may be easily recognized and differentiated from other diseases. Therefore, knowledge of the various hallmarks of different tauopathies enables neuropathologists not only to diagnose specific disorders but also to recognize different lesions co-occuring in one and the same brain and to assess the impact of individual lesions on brain dysfunction.

The present paper is based on the authors' experiences obtained from a total of nearly 130 AgD cases examined between 1994 and 1999 at the Institute of Pathology, Basel University. It is aimed at acquainting neuropathologists but also clinicians with the morphological hallmark lesions of AgD. Moreover we will focus on recent genetic and clinical findings obtained in AgD cases.

2. NEUROPATHOLOGY OF AGD – CORE LESIONS

2.1 Gross aspect

Upon gross examination, the brains of AgD patients appear virtually unchanged or show only mild diffuse or frontotemporal cortical atrophy. In particular, there is no obvious atrophy of the hippocampus formation and/or the amygdaloid complex, and there is no enlargement of the temporal portion of the lateral ventricles[7,9,15]. In our series of AgD cases the mean brain weight was 1203g (1254g for men and 1151g for female patients). These values only slightly differed from those reported in octogenarian subjects without a cognitive decline[16]. It was of interest that the mean brain weight for demented AgD patients significantly differed from that of non-demented subjects with ArGs (1180g and 1255g, respectively, student t-test

$p<0.004$; for a definition of demented and non-demented AgD patients see below). Therefore, with the help of a previously published computer program which combines the independent variables, age and body height, in a nomogram for the calculation of brain weight (PathoStat, for details see[16]) we determined the corresponding percentiles on which the brain weight of the AgD cases were located. The mean percentile value for all AgD cases was 49.28 (range 0.94 - 99.37) which corresponds to the value of mentally unimpaired subjects[16]. In contrast to the non-corrected brain weight, no significant difference was found when we compared the mean percentile values of demented and non-demented AgD subjects (mean value for demented AgD patients: 47.02; non-demented AgD patients: 54.93, $p=0.2$).

The mean brain weight from a series of 30 autopsy confirmed AD patients was 1098g, and significantly differed from both, mean brain weight for demented and non-demented AgD patients ($p<0.01$ and $p<0.0001$, respectively). Similar results were obtained when we compared the corresponding percentile values between AD (mean value: 23.4, range 0.8 - 68.7) and demented and non-demented AgD subjects (for both $p<0.0001$).

2.2 Histology

2.2.1 Argyrophilic grains (ArGs)

ArGs constitute the most important histopathological hallmark lesion of AgD (Figure 1a-c, Table 1). These oval, spindle-shaped or comma-like lesions can best be detected in routine paraffin embedded tissue by the use of conventional silver methods (e.g., Bielschowsky or Bodian stain), but are more easily recognized in tissue sections processed with the Gallyas silver iodide technique[17]. However, in AgD only a fraction of the emerging abnormal intracellular deposits consists of argyrophilic fibrils. Silver staining method therefore is adequate only for diagnostic purposes. The bulk of AgD-pathology (ArGs, pretangle neurons, glial changes) is visible only by immunoreactions for abnormally phosphorylated tau protein (see below).

ArGs are found in both cortical and subcortical structures. Among cortical areas the transentorhinal and entorhinal cortices which are located in anteromedial portions of the temporal lobe are most susceptible to the development of ArGs[7,15,18]. The brunt of ArG change is found in the upper pre-ß layer of the entorhinal and transentorhinal cortex, although scattered ArGs are also present within the deep pri-α layer and the cellular islands of layer pre-α. The highest density of ArGs is usually found close to the transentorhinal-entorhinal border. From the amygdala (see below) and the temporal allocortex ArGs typically spread out into layer III of the temporal neocortex. ArGs are absent or only sparse in the anterobasal portions of the insula, as well as in the temporopolar and frontoorbital neocortex. ArGs are almost absent in other neocortical regions. ArGs are abundantly found in the external and internal pyramidal layers of the first Ammon's horn sector (CA1) (Figure 1a) with a striking decrease in density towards both the stratum radiatum-lacunosum-moleculare and the stratum oriens. Beside sector CA1, the wedge-shaped prosubiculum usually exhibit the highest ArG density in the hippocampus region. In contrast, the subiculum proper only contains a small number of ArGs. Few ArGs may be found in sectors CA2-4 and the parasubiculum in some cases.

At the subcortical level ArGs are densely distributed throughout the basolateral nuclear group of the amygdaloid complex. ArGs are also found in other subnuclei of the amygdala however, at lower densities. High densities of ArGs are also present in the hypothalamic lateral tuberal nucleus, whereas the adjoining hypothalamic nuclei and the entire thalamus appear nearly devoid of lesions[19]. On rare occasions, ArGs may be found in basal portions of the claustrum but the striatum and the magnocellular nuclei of the basal forebrain remain consistently free of grains. The lower brain stem occasionally displays a few ArGs in the periaqueductal gray matter, anterior raphe nuclei, locus coeruleus, and dorsal vagus area. In the vast majority of our cases diencephalic and mesencephalic components of the basal ganglia as well as the globus pallidus, cerebellum, pons, inferior olive, and spinal cord do not contain ArGs.

By using the Gallyas silver iodide technique ArGs are loosely scattered throughout the neuropil (Figure 1a) and typically display a spindle-shaped sometimes slightly bent body with a long axis extending up to 9 μm and a diameter reaching approximately 4 μm. Variations in shape include straight or kinked rod- or drumstick-like structures, often with tiny excrescences along the surface. Occasionally, spine-like protrusions and knobby or filiform appendages are observed arising from both cone-shaped poles (Figure 1b, c). Quite often, grains are arranged in short rows, where individual grains appear interconnected by tiny thread-like structures.

ArGs can be easily distinguished from neurofibrillary lesions (NFL) of AD. They are much smaller than neurofibrillary tangles (NFTs) and are never seen in neuronal cell perikarya (see below). Extracellular "ghost" or "tombstone" tangles are larger in size, and they show much less argyrophilia than ArGs[20,21]. Neuropil threads (NThs) are slender filiform structures without spindle-shaped thickenings and are more variable in size and shape than the sturdy ArGs which exhibit only little variation[22]. Although ArGs are distributed evenly throughout the neuropil, clusters of grains may be observed in some cases which may - at a first glance - resemble neuritic plaques (NPs) of AD or astrocytic plaques of CBD. However, dystrophic neurites of NPs vary greatly in size and shape when compared to the much more uniform ArGs, and the diameter of both NPs and astrocytic plaques generally exceed that of ArGs clusters. In addition, ß-amyloid deposits are not encountered in clustered ArGs. ArGs can also be easily distinguished from tufted astrocytes of PSP. Pick bodies (PBs) are much larger than ArGs and are exclusively found in neuronal cell perikarya. Furthermore, PBs, although argyrophilic, are not stained with the Gallyas silver impregnation technique[10].

At the ultrastructural level ArGs also differ from NFL observed in other tauopathies. ArGs consist of aggregates of 9 to 18 nm filaments[14,18,23] and of bundles of 25 nm smooth tubules[24,25]. Paired helical filaments of AD or twisted ribbons reported in some FTDP-17 families[5] are not a component of ArGs.

2.2.2 Coiled bodies

Coiled bodies are conspicuous curvaceous whip-like often branched oligodendroglial inclusions which are in close vicinity to the cell nucleus (Figure 1d). By electron microscopy, they consist of straight 10-13 nm filaments or tubular structures[23,24].

Coiled bodies may be found in various amounts in several tauopathies (e.g. PSP, CBD, PiD, AD and FTDP-17), and are therefore considered as a non-specific finding[26,27]. However, in AgD, coiled bodies represent a consistent finding in the white matter close to cortical areas and subcortical nuclei rich in ArGs, and coiled bodies may also be encountered in deep layers of affected cortical regions. Therefore, in addition to ArGs, we consider coiled bodies as an essential hallmark lesion of AgD (Table 1). Coiled bodies in AgD typically accompany the perforant path, as well as the alveus and fornix throughout their entire course. However, it is notable that only a small subset of oligodendroglial cells within a given volume of white matter exhibit the pathological change[7,19].

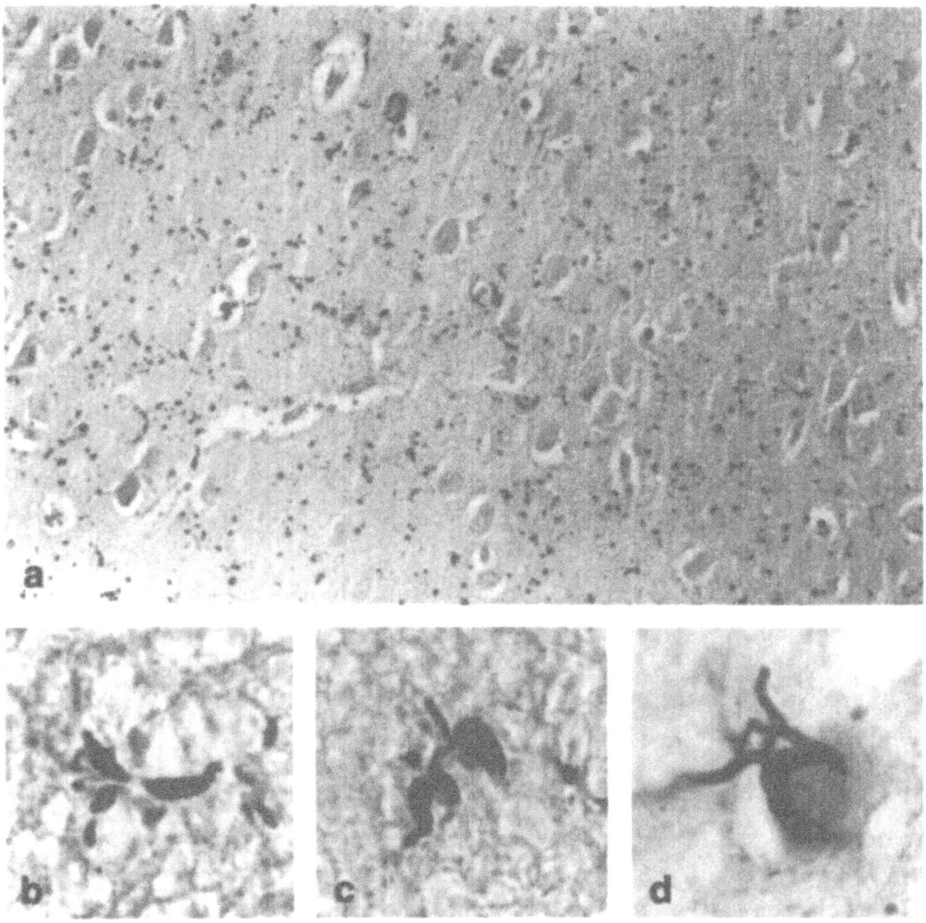

Figure 1. **a.** Argyrophilic grains in the intercellular grey matter of the hippocampus (sector CA1). Note the complete absence of neurofibrillary changes and neuropil threads in this area of the cortex. **b,c,** Argyrophilic grains sometimes show saccular or filiform protrusions. **d.** Subcortical white matter, hippocampus region. One coiled body surrounding the cell nucleus of an oligodendrocyte (a-d: Gallyas stain, a x280; b-d x875).

2.3 Tau protein pathology in neurons and glial cells

Several studies have shown that the main protein constituent of ArGs and coiled bodies is the abnormally and hyperphosphorylated tau protein (reviewed in[15]). In extension of these studies we have recently shown that tau protein in ArGs and coiled bodies share a high number of phosphorylated sites with NFTs and NThs of AD (PHF-tau), among them Thr 181, Ser 202, Thr 205, Thr212, Ser214, Thr 231, Ser 235, Ser 396 and Ser 404. Our finding that a number of N- and C-terminal sites and the first and second N-terminal 29 amino acid inserts of tau are also recognized by a panel of non phosphorylation-dependent antibodies indicates that full length tau is contained in ArGs and coiled bodies[18].

A notable exception is antibody 12E8 directed against phosphorylated Ser 262 and/or Ser 356 of tau[28]. While NFTs in AD are strongly stained, this antibody consistently fails to stain ArGs and coiled bodies of AgD[18]. Ser 262 is located in the first microtubule-binding repeat of tau and has been shown to be phosphorylated in a fraction of PHF-tau[29]. Phosphorylation of Ser 262 has also been reported to regulate the binding of tau to microtubules[30]. However, 12E8 has recently been shown to recognize not only PHF-tau, but also a significant fraction of fetal rat brain tau as well as a small fraction of adult rat brain tau and biopsy-derived adult human brain tau[28]. Moreover, phosphorylation of Ser 262 only poorly affects tau binding to microtubules, suggesting that phosphorylation of multiple residues is needed to disrupt interactions between tau and microtubules[28]. Hyperphosphorylation of Ser 262 might possibly be a late event that occurs only in highly aggregated and compacted neurofibrillary lesions, as observed in conditions such as AD and Niemann-Pick disease type C[31]. Such a high degree of aggregation and condensation is probably never reached in ArGs or coiled bodies of AgD, as suggested by the lack of PHFs in these lesions (see above). We have also shown that hyperphosphorylated tau protein of Pick bodies fails to react with antibody 12E8[10], an observation recently confirmed by Delacourte and coworkers[32].

By using tau-immunohistochemistry (e.g. antibody AT8[33]) a main finding in AgD cases consists of a widespread hyperphosphorylation of tau protein in the somatodendritic domain of neurons. Tau hyperphosphorylation in neurons is found in areas rich in silver grains, such as sector CA1 of the hippocampus, the transentorhinal/entorhinal cortex and the amygdaloid nucleus. In sector CA1 one may find up to 80% of pyramidal cells exhibiting diffuse AT8 staining of their cell bodies and dendrites. Because this constitutes a highly characteristic finding for AgD we consider it as one of the core lesions yet not essential for routine diagnosis (Table 1). The intraneuronal distribution of hyperphosphorylated tau in AgD looks very similar to the "stage 0 tangles" of Bancher et al.[20] and the "group 1 neurons" of Braak et al.[21], which in AD have been considered as possible precursor stages of neurofibrillary degeneration (pretangle neurons; see below). In AgD all of the neuronal types prone to accumulate hyperphosphorylated tau protein belong to the class of projection neurons and include constituents of the transentorhinal and entorhinal layers pre-α and pri-α, the principal cells of the hypothalamic lateral tuberal nucleus, the projection cells in both the central nucleus and the basolateral nuclei of the amygdala, the external and internal pyramidal cells in sector CA1, as well as the hilar mossy cells and the granule cells of the fascia dentata[7,18].

Recent studies showed that a phosphorylation-independent interaction between full-length tau and sulfated glycosaminoglycans, such as heparin and heparan sulfate, leads to the formation of filaments which closely resemble those found in AD brains[34]. Immunohistochemical studies have shown that heparan sulfate and hyperphosphorylated tau colocalize in pretangle neurons in AD brains. Moreover, in some nerve cells, the heparan sulfate staining was found to be more extensive than the tau staining and some tau-negative nerve cells were heparan sulfate-positive, thus suggesting that the accumulation of heparan sulfate precedes hyperphosphorylation of tau[34]. We recently also obtained a very similar finding in pretangle neurons of our AgD cases[35]. These findings suggest that an interaction between tau protein and sulfated glycosaminoglycans may also play an important role in inducing both the hyperphosphorylation of tau and the formation of ArGs in AgD.

By using combined AT8/Gallyas staining we have recently shown that at least a subset of ArGs are formed within dendrites and dendritic side branches but not in perikarya of neurons containing hyperphosphorylated tau protein[36]. This finding is in accordance with previous studies suggesting a dendritic origin of ArGs[19,24]. In addition, the morphological appearance of neurons containing ArGs, and especially of their dendritic branches, strongly suggests an ongoing process of dendritic shrinkage leading to progressive shortening and loss of dendritic ramifications (see below).

As aforementioned the intraneuronal distribution of hyperphosphorylated tau in AgD looks very similar to pretangle neurons in AD. However, in contrast to AD where the stage of pretangle neurons is obviously followed by the formation of neurofibrillary lesions in neuronal perikarya and dendrites, hyperphosphorylation of tau in AgD does not lead to the formation of neurofibrillary lesions but instead to the formation of ArGs in dendrites and dendritic side branches[36]. This also differs from PiD where abnormal tau is mainly somatoaxonal with formation of Pick bodies in neuronal perikarya[10].

Astrocytes have been considered not or only occasionally affected in AgD[7,15,26,27]. However, we recently reported the finding of non-argyrophilic astrocytes expressing hyperphosphorylated tau epitopes as a constant feature in anterior limbic structures like the amygdala and anterior entorhinal cortex in AgD cases[37]. This observation points to a more widespread tau pathology of the glial cell population in AgD than previously admitted. Argyrophilic tau-positive inclusions in glial cells, also termed glial fibrillary tangles (GFTs), have been described in many neurodegenerative disorders in which there is also tangle pathology in neurons. Morphologically, glial tangle pathology includes tufted astrocytes, thorn-shaped astrocytes, astrocytic plaques, coiled bodies (see above) as well as interfascicular and white matter threads (for review see[4,26,27]). None of these morphological subtypes seem to be specific for a particular disease, and various subtypes of GFTs may be present in one particular disorder, such as in PSP, CBD or PiD.

The tau-positive non-argyrophilic astrocytes in AgD share some features with tufted astrocytes mainly observed in basal ganglia and cortical regions of PSP, with tau-positive astrocytes in the neocortex and subcortical white matter of PiD and with astrocytic plaques of CBD. In AgD, AT8-immunostaining is found in all cell compartments of affected astrocytes which contrasts with astrocytic plaques of CBD where tau protein is mainly concentrated in distal cell processes, and with tufted astrocytes of PSP where tau immunoreactivity is mainly confined to proximal and distal processes with relative spearing of the cell soma[4]. Moreover, in contrast to all these astrocytic inclusions, there are no argyrophilic GFTs in tau-immunoreactive astrocytes of AgD. Thus, tau-expressing astrocytes in AgD, like their neuronal counterpart, are likely to remain in a state of tau hyperphosphorylation without tendency to fibrillary aggregation of tau.

3. NEUROPATHOLOGY OF AGD – ASSOCIATED LESIONS

3.1 Lesions of the Alzheimer-type

The main reason why AgD is questioned to represent a distinct disease entity is the observation that most of the reported AgD cases have associated changes of the AD-type, e.g. neurofibrillary lesions (Table 1). Some authors therefore consider AgD as a variant of AD[38]. Closer inspection, however, reveals notable differences between AD-type changes in AgD and AD, which strongly argue against the former hypothesis.

Two large series of AgD cases, in total 238 cases[7,39], revealed extension of neurofibrillary changes in the shape of NFTs and NThs corresponding to early (entorhinal and limbic) Braak stages[40]. This extension of NFL is a frequent finding in aged subjects and is generally not associated with a cognitive decline[40,41]. In both series the most frequently co-occurring NFT stage was Braak stage II (n = 94) followed by stages III (n = 65), I (n = 60) and IV (n = 11). Only 4 cases were devoid of neurofibrillary changes whereas 4 showed concomitant fully-developed AD at stage V. Obviously, pretangle limbic neurons which are abundantly found in AgD do not further evolve to a stage of tangle bearing cells which is in sharp contrast to what is observed in AD (see above).

When using appropriate silver staining techniques and/or Aß immunohistochemistry, senile plaques (SP) are found in about two thirds of AgD cases, whereas the remaining cases are completely devoid of SP[7,39]. It is notable that the same proportion of AgD patients with and without SP are found in demented AgD patients and in those without a cognitive decline (see below)[39] making it very unlikely that Aß deposits cause dementia in AgD. Using stereological principles we have shown that the average total amyloid (Aß) plaque load in AgD is significantly lower than in AD. Moreover the regional destribution of SP and the proportion of diffuse versus primitive or mature plaques in AgD are similar to values of SP reported in non-demented elderly, and is significantly different from AD. Similarly the immunocytochemical profile of the Aß deposition in AgD resembles that of cognitively normal elderly rather than that of AD[39].

3.2 Ballooned cells

Ballooned neurons (BNs) are characterized by a swollen pale-staining cytoplasm and a large eccentrically located nucleus. They have been described in the cerebral cortex in various neurodegenerative disorders, among them PiD, CBD, PSP, CJD and AD[42,43]. As an inconsistent finding few to moderate numbers of cortical BNs may be also encountered in layers V and VI of basal temporal neocortical areas in AgD cases[7,44,45]. However, in contrast to cortical regions, BNs are a constant feature of amygdaloid nuclei in patients with AgD[45] (Table 1). BNs in AgD are strongly labelled with antibodies against αB-crystallin, a feature they share with BNs in other neurodegenerative disorders[46]. Moreover they also stain with antibodies against phosphorylated neurofilament and phosphorylated tau. However, in contrast tau immunoreactive non-ballooned neurons (pretangle neurons), ArGs and coiled bodies remain consistently unstained with αB-crystallin antibodies[45]. These findings point to two different pathological mechanisms involved in limbic neurons in AgD: (i) a neuronal pathology encompassing hyperphosphorylation of tau in all cell compartments linked to the formation of ArGs without cell ballooning or αB-crystallin expression; (ii) a pathology of nerve cells, whose most distinctive feature is cell ballooning associated with hyperphosphorylation of tau and αB-crystallin expression. This type of nerve cell change is obviously not related to ArGs formation.

3.3 Other associated lesions

In some AgD cases one may find a prominent superficial laminar spongiosis in cortical areas rich in ArGs like the entorhinal-transentorhinal cortex and the adjacent temporal neocortex[44]. A moderate gliosis may be present in the entorhinal cortex, posterior parahippocampal gyrus and the amygdala as well[44]. Moreover, Yamada et al.[47] reported on two AgD cases with extensive subcortical gliosis without significant neuronal loss. However, in our own series of AgD cases we do not find such an extent of gliosis in cortical or subcortical structures. Based on the literature and our own experience it seems likely that both findings – cortical spongiosis and subcortical gliosis – represent inconsistent findings and therefore are not essential to the diagnosis of AgD (Table 1).

I. Core lesions of AgD
Essential for diagnosis
a) Argyrophilic grains
Consistent features but not essential for diagnosis
b) Coiled bodies
c) Abundant non-argyrophilic limbic projection neurons containing hyperphosphorylated tau protein in all cell compartments

II. AgD associated lesions - frequent findings but not essential for diagnosis
a) Ballooned neurons (frequent finding in amygdala and layers V and VI of basal temporal neocortical areas)
b) Non-argyrophilic tau immunoreactive astrocytes (frequent finding in amygdala and entorhinal and transentorhinal cortex)
c) Associated lesions of the Alzheimer-type
- Neurofibrillary lesions corresponding to Braak stages I-III (frequent finding)
- Few senile plaques, mainly of the diffuse type (two thirds of cases)

III. AgD associated lesions - rare findings
a) Superficial laminar spongiosis (layers II-III of basal temporal neocortical areas)
b) Cortical and subcortical gliosis (entorhinal and transentorhinal cortex, posterior parahippocampal gyrus)

Table 1. Argyrophilic grain disease: neuropathological features

3.4 AgD and co-occurrence of other neurodegenerative disorders

ArGs have been reported in a number of neurodegenerative disorders, among them PiD[24,44], PSP[24,25,44], CBD[24,44], PD[7,14,18,24,48], dementia with Lewy bodies[24,44], tangle predominant form of senile dementia ("tangle only dementia")[49,50], Creutzfeldt-Jakob disease[51] and multiple system atrophy[44].

AgD has therefore been questioned to be a separate disease entity, and some authors suggested to classify AgD as a variant of the lobar atrophy complex (Pick's complex) or of PSP[44]. However, in the mentioned disorders ArGs are an inconsistent feature, found in occasional cases. Moreover, although not regularly present, ArGs constitute a frequent finding in brains from old and very old subjects (see below), and it therefore does not surprise that some ArGs may be found as an incidental feature in various neurodegenerative disorders in old age.

"Tangle only dementia" denotes a relatively rare subset of demented elderly patients who share a uniform neuropathological picture consisting of moderate cerebral atrophy and presence of various numbers of NFL in allocortical areas of the inferomedial temporal lobe (hippocampus, subiculum, entorhinal region) and the amygdala, corresponding to Braak stages III and IV, with absence or extreme scarcity of amyloid deposits and neuritic plaques[49,50,52]. ArGs were reported to be associated with 20 - 66% of "tangle only dementia" cases[44,50]. Therefore AgD and tangle only dementia have been claimed to be one and the same pathological condition. Moreover the apolipoprotein E ε4 allele was not found to constitute a risk factor for the development of tangle only dementia[53,54]. Intriguingly, a very similar finding was also reported recently in two independent studies for AgD[55,56] (see below). Further studies are now aimed to elucidate morphological and genetic similarities and/or differences between tangle only dementia and AgD (Tolnay et al. manuscript in preparation).

4. EPIDEMIOLOGY

4.1 Prevalence of AgD

Although ArGs constitute a well defined morphological change, easy to recognize with appropriate silver stain at postmortem examination (see above), only few studies are available on the prevalence of AgD in the elderly population. In their original report, Braak and Braak found ArGs as the main histopathological feature in 28 out of 80 brains obtained at autopsy from old-aged demented subjects, thus suggesting a very high prevalence of grains among old demented patients[14]. However, ArGs constituted the most salient morphological change in only 10, while in 16 subjects ArGs were associated with NFTs, NThs and various numbers of neuritic plaques. Itagaki et al.[23] found AgD in only one case among 33 consecutive autopsies of old demented subjects without cerebrovascular pathology. Davis et al.[57]

found ArGs in 14 (23%) brains among 59 aged cognitively normal subjects (42, 43).

In a recent study Braak and Braak found 125 (5%) AgD cases among 2261 non-selected autopsies, all but 5 of which were over 60 years of age[7]. A cognitive decline was reported in about 20% of AgD cases in Braak's series. This correlates with data reported by Martinez-Lage and Munoz who reported 17 (5,6%) AgD cases in their study of 300 consecutive autopsies of patients older than 30 years, or 11,7% when only subjects older than 65 were considered[44]. However, among these 17 cases 6 had other neuropathological diagnoses of neurodegenerative disorders. Among the remaining 11 subjects 2 (18%) had shown evidence of mental impairment[44]. In a previous study of 301 consecutive autopsies of patients older than 65, we found 29 (9,6%) cases with AgD[18,58]. Among these 29 patients only 14 (48%) had presented progressive cognitive decline during the last year of their life. In most cases, ArGs were associated with other morphological changes, among them NFT and senile plaques. However, changes of the Alzheimer type were not found to be more severe in demented patients, when compared with subjects without cognitive decline[18].

From all these data it can be concluded that AgD-changes do not constitute an obligate finding in cognitively unimpaired subjects and therefore cannot be regarded as true age-related changes of the human brain. Nevertheless, the prevalence of AgD cases clearly increases with advancing age[7].

4.2 ArGs in demented and non-demented subjects

As mentioned earlier AD-type lesions do obviously not cause dementia in AgD. One might therefore hypothesize that ArGs *per se* are causal for the cognitive decline observed in some AgD patients. We recently addressed this question by looking more precisely at the distribution and density of ArGs in well documented demented and non-demented cases. It is suggested from this study that there is a significantly more widespread rostrocaudal extension of ArG change in the limbic area in subjects with documented intellectual decline than in cognitively unimpaired subjects[59]. The caudal part of CA1 is significantly more often affected in AgD cases with dementia whereas no differences in distribution and density of ArGs is found in the rostral CA1. In addition ArGs are significantly more widespread in the entorhinal/transenthorhinal region and the parahippocampal cortex of demented patients[59]. None of our AgD cases showed an exclusive involvement of the the posterior hippocampal region without an involvement of the anterior region. This suggests that early subclinical changes in AgD might begin in the rostral hippocampus formation and/or in the most anterior entorhinal cortex or amygdaloid complex. Thus, together with the previous finding that ArGs are formed within dendritic branches of pretangle neurons and that we observed a progressive shrinkage and loss of dendritic branches bearing ArGs one might speculate that one possible explanation for the dementia observed in some subjects with AgD is that there is a more widespread loss of postsynaptic structures, including synaptic contacts, throughout the hippocampus-entorhinal/parahippocampal complex, and probably the amygdaloid nuclei[36,59].

5. APOE GENOTYPE

The ε4 allele of the ApoE gene is a strong susceptibility factor for the late-onset and sporadic forms of AD, with a roughly three-fold increase in frequency among AD patients compared to age matched controls (for a review see[60]). Today, ApoE genotyping is recommended as an adjunct test in the diagnostic workup of an individual with suspected AD where the finding of an ε4 allele adds a small percentage of confidence to the clinical diagnosis[61]. In addition to AD, an association with the ε4 allele has also been reported for other tauopathies like PiD, PSP and CBD[62]. Recently, the ApoE genotypes have been analysed in two independent studies for subjects with ArGs. A very low ApoE ε4 allele frequency is found in AgD, which significantly differs from the ε4 allele frequencies in AD, but not from age-matched control subjects[55,56]. In one of these studies a significantly higher incidence of the ApoE ε2 allele is reported in AgD cases than in AD and age-matched control patients[56]. However, extended clinicopathological studies will be necessary to resolve the question whether a subpopulation of eldery individuals with the ApoE ε2 allele is indeed at higher risk for developing cognitive dysfunction due to AgD-related changes. At present, it is suggested from both studies that the ApoE ε4 allele does not constitute a risk factor for the development of AgD[8,55,56].

6. CLINICAL ASPECTS OF AGD

AgD shows a significant correlation with advancing age[7,9]. Thus old age can be considered as the main risk factor for AgD. To the best of our knowledge no familial and presenile case have been reported so far. ArGs may or may not be associated with a cognitive decline (see above). In the original paper by Braak and Braak[14] all demented patients were reported to suffer from adult-onset progressive mental deterioration. During the course of the disease patients became anxious, restless, often depressed and emotionally shallow. Many of them showed incontinence and cachexia[14].

At present only little is known about clinical aspects and especially about early (precognitive) symptoms of AgD patients. Quite detailed clinical data have been provided by Itagaki et al.[23] and Yamada et al.[47] in their account on two patients with AgD. However, it is uncertain if their patients represent "typical" AgD cases because postmortem examination of the brains, in addition to ArGs, also revealed extensive subcortical pathology (astrocytic gliosis in striatum, amygdala, thalamus, substantia nigra). The same applies for the five "AgD" patients reported by Masliah et al.[25]. In addition to ArGs these authors also found widespread tangles and threads in a topographical distribution suggestive of progressive supranuclear palsy. Thus, they termed their cases either as "late onset dementia with argyrophilic grains and subcortical tangles" or as "atypical progressive supranuclear palsy"[25].

Ikeda et al.[24] reported on two patients with ArGs as the main histopathological finding. Both patients (2 men, 77 and 84 years old, respectively) had quite similar clinical histories. Interestingly both patients had been refered to a mental hospital because of emotional disorder, characterised by paroxysmal excitement and violence, rather than a dementia which was reported to be only mild. The duration of the clinical course was 11 and 20 months, respectively[24]. Some of these findings are in accordance with those reported by Braak and coworkers[7,63]. It is suggested from these studies that cognitive impairment does not constitute an early event in AgD patients. Instead, in contrast to what is observed in AD patients, it seems likely that in AgD behavioural disturbances (paroxysmal excitement and violence, inappropriate social conduct, personality changes, aggressivness, restlessness, loss of personal awareness) precede memory failure and cognitive decline[7,63,own unpublished observation].

In late stages of the disease process, when dementia is already present, AgD may become clinically indistinguishable from AD. At this stage, brain imaging of demented patients could be of help in the diagnostic process. Volumetric magnetic resonance imaging (MRI) measurements of the hippocampus and entorhinal cortex can distinguish AD patients reliably from cognitively unimpaired elderly subjects (for a recent study see[64]). However, it became clear with additional research that hippocampal atrophy is not specific for AD, but may be also found to some degree in other dementing disorders like Parkinson's disease (with or without dementia), dementia with Lewy bodies as well as dementia of frontal type[65]. It is suggested from these studies that differential patterns of brain atrophy might help in the diagnostic process of dementia cases (for review see[65]).

As mentioned earlier it became apparent from several postmortem studies that most brains from AgD patients appear virtually unchanged or show only mild diffuse or frontotemporal cortical atrophy. Most notable, there is no obvious atrophy of the hippocampus formation, the entorhinal cortex and/or the amygdaloid complex. Therefore, volumetric MRI measurements of both the entorhinal cortex and hippocampus might be of help in distinguishing (demented) AgD patients from AD patients.

Further retrospective and prospective clinicopathological studies are now aimed to document in detail specific features which would permit an (early) clinical diagnosis of AgD.

7. CONCLUSIONS

In conclusion recent studies strongly suggest that argyrophilic grain disease represents a dementing disorder of old age more frequent than previously thought. Moreover it seems very likely that argyrophilic grain disease is indeed a distinct disease entity separate from other tauopathies, e.g. from Alzheimer's disease by both morphological and genetic criteria. Preliminary data from ongoing clinical studies further suggest that argyrophilic grain disease may also be distinguished clinically from Alzheimer's disease, at least in early stages of the disease process. The correct diagnosis of AgD, or co-occurrence of AgD with other neurodegenerative disorders will reduce considerably the number of enigmatic and unexplained cases of late-onset dementia.

REFERENCES

1. Goedert M, Trojanowski JQ, Lee VM-L (1997) The neurofibrillary pathology of Alzheimer's disease. In: The molecular and genetic basis of neurological disease (Rosenberg RN, Prusiner SB, Di Mauro S, Barchi RL, eds), pp. 613-622: Butterworth-Heinemann, Boston.
2. Selkoe DJ (1997) Cellular and molecular biology of the beta-amyloid precursor protein and Alzheimer's disease. In: The molecular and genetic basis of neurological disease (Rosenberg RN, Prusiner SB, Di Mauro S, Barchi RL, eds), pp. 601-611: Butterworth-Heinemann, Boston.
3. Spillantini MG, Goedert M (1998) Tau protein pathology in neurodegenerative disease. Trends in Neurosci 21: 428-433.
4. Tolnay M, Probst A (1999) Review: Tau protein pathology in Alzheimer's disease and related disorders. Neuropathol Appl Neurobiol 25: 171-187.
5. Spillantini MG, Van Swieten JC, Goedert M (2000) Tau gene mutations in frontotemporal dementia and parkinsonism linked to chromosome 17 (FTDP-17). Neurogenetics 2: 193-205.
6. Buée L, Delacourte A (1999) Comparative biochemistry of tau in progressive supranuclear palsy, corticobasal degeneration, FTDP-17 and Pick's disease. Brain Pathol 9: 681-693.

7. Braak H, Braak E (1998) Argyrophilic grain disease: frequency of occurrence in different age categories and neuropathological diagnostic criteria. J Neural Transm 105: 801-819.
8. Tolnay M, Probst A (1999) Argyrophilic grain disease: a neurodegenerative disorder distinct from Alzheimer's disease. In: Alzheimer's disease and related disorders: etiology, pathogenesis and therapeutics (Iqbal K, Swaab DF, Winblad B, Wisniewski HM, eds), pp 537-543: John Wiley and Sons, Ltd., Sussex.
9. Probst A, Tolnay M (2000) La maladie des grains argyrophiles: une cause fréquente mais encore largement méconnue de démence chez les personnes âgées. Rev Neurol: in press
10. Probst A, Tolnay M, Langui D et al. (1996) Pick's disease: hyperphosphorylated tau protein segregates to the somatoaxonal compartment. Acta Neuropathol 92: 588-596.
11. Feany MB, DW Dickson (1995) Widespread cytoskeletal pathology characterizes corticobasal degeneration. Am J Pathol 146: 1388-1396.
12. Komori TN, Arai H, Oda H (1998) Astrocytic plaques and tufts of abnormal fibres do not coexist in corticobasal degeneration and progressive supranuclear palsy. Acta Neuropathol 96: 401-408
13. Braak, H, Braak E (1987) Argyrophilic grains: characteristic pathology of cerebral cortex in cases of adult onset dementia without Alzheimer changes. Neurosci Lett 76: 124-127.
14. Braak H, Braak E (1989) Cortical and subcortical argyrophilic grains characterize a disease associated with adult onset dementia. Neuropathol Appl Neurobiol 15: 13-26.
15. Jellinger K (1998) Dementia with grains (argyrophilic grain disease). Brain Pathol 8: 377-386.
16. Hartmann P, Ramseier A, Gudat F et al. (1994) Normal weight of the brain in adults, referred to age, sex, height and body weight. Pathologe 15: 165-170.
17. Gallyas F (1971) Silver staining of Alzheimer's neurofibrillary changes by means of physical development. Acta Morphol Acad Sci Hung 19: 1-8.
18. Tolnay M, Spillantini MG, Goedert M et al. (1997) Argyrophilic grain disease: widespread hyperphosphorylation of tau protein in limbic neurons. Acta Neuropathol 93: 477-484.
19. Schultz C, Koppers D, Sassin I et al. (1998) Cytoskeletal alterations in the human tuberal hypothalamus related to argyrophilic grain disease. Acta Neuropathol 96: 596-602
20. Bancher C, Brunner C, Lassmann H et al. (1989) Accumulation of abnormally phosphorylated tau precedes the formation of neurofibrillary tangles in Alzheimer's disease. Brain Res 477: 90-99.
21. Braak E, Braak H, Mandelkow EM (1994) A sequence of cytoskeleton changes related to the formation of neurofibrillary tangles and neuropil threads. Acta Neuropathol 87: 554-567.
22. Braak H, Braak E, Grundke-Iqbal I et al. (1986) Occurence of neuropil threads in the senile human brain and in Alzheimer's disease: a third location of paired helical filaments outside of neurofibrillary tangles and neuritic plaques. Neurosci Lett 65: 351-355.
23. Itagaki S, McGeer PL, Akiyama H et al. (1989) A case of adult-onset dementia with argyrophilic grains. Ann Neurol 26: 685-689.
24. Ikeda K, Akiyama H, Kondo H et al. (1995) A study of dementia with argyrophilic grains. Possible cytoskeletal abnormality in dendrospinal portion of neurons and oligodendroglia. Acta Neuropathol 89: 409-414.

25. Masliah E, Hansen LA, Quijata S et al. (1991) Late onset dementia with argyrophilic grains and subcortical tangles or atypical progressive supranuclear palsy? Ann Neurol 29: 389-396.
26. Chin SSM, Goldman EJ (1996) Glial inclusions in CNS degenerative diseases. J Neuropathol Exp Neurol 55: 499-508.
27. Ikeda K, Akiyama H, Arai T et al. (1998) Glial tau pathology in neurodegenerative diseases: their nature and comparison with neuronal tangles. Neurobiol Aging 19: 85-91.
28. Seubert P, Mawal-Dewan M, Barbour R et al. (1995) Detection of phosphorylated Ser 262 in fetal tau, adult tau and paired helical filament-tau. J Biol Chem 270: 18917-18922.
29. Hasegawa M, Morishima-Kawashima M, Takio K et al. (1992) Protein sequence and mass spectrometric analyses of tau in the Alzheimer's disease brain. J Biol Chem 267: 17047-17054.
30. Biernat J, Gustke N, Drewes G et al. (1993) Phosphorylation of Ser 262 strongly reduces binding of tau to microtubule: distinction between PHF-like immuno-reactivity and microtubule binding. Neuron 11: 153-163.
31. Auer IA, Schmidt ML, Lee VM-Y et al. (1995) Paired helical filament tau (PHFtau) in Niemann-Pick type C disease is similar to PHFtau in Alzheimer's disease. Acta Neuropathol 90: 547-551.
32. Delacourte A, Sergeant N, Wattez A et al. (1998) Vulnerable neuronal subsets in Alzheimer's and Pick's disease are distinguished by their tau isoform distribution and phosphorylation. Ann Neurol 43: 193-204.
33. Goedert M, Jakes R, Vanmechelen E (1995) Monoclonal antibody AT8 recognizes tau protein phosphorylated at both serine 202 and threonine 205. Neurosci Lett 189: 167-170.
34. Goedert M, Jakes R, Spillantini MG et al. (1996) Sulphated glycosaminoglycans induce assembly of microtubule-associated protein tau into Alzheimer-like filaments. Nature 383: 550-553.
35. Spillantini MG, Tolnay M, Love S et al. (1999) Microtubule associated protein tau, heparan sulphate and α-synuclein in several neurodegenerative diseases with dementia. Acta Neuropathol 97: 585-594.
36. Tolnay M, Mistl C, Ipsen S et al. (1998) Argyrophilic grains of Braak: occurrence in dendrites of neurons containing hyperphosphorylated tau protein. Neuropathol Appl Neurobiol 24: 53-59.
37. Botez G, Probst A, Ipsen S et al. (1999) Astrocytes expressing hyperphosphorylated tau protein without glial fibrillary tangles in argyrophilic grain disease. Acta Neuropathol 98: 251-256.
38. Cras P, Perry G (1991) Dementia with argyrophilic grains. Ann Neurol 30: 853-854.
39. Tolnay M, Calhoun M, Pham HC et al. (1999) Low amyloid (Aß) plaque load and relative predominance of diffuse plaques distinguish argyrophilic grain disease from Alzheimer's disease. Neuropathol Appl Neurobiol 25: 295-305.
40. Braak H, Braak E (1991) Neuropathological stageing of Alzheimer-related changes. Acta Neuropathol 82: 239-259.
41. Braak H, Braak E (1997) Frequency of stages of Alzheimer-related lesions in different age categories. Neurobiol Aging 18: 351-357.
42. Dickson DW, Yen SH, Suzuki KI et al. (1986) Ballooned neurons in select neurodegenerative diseases contain phosphorylated neurofilament epitopes. Acta Neuropathol 71: 216-223.

43. Mori H, Oda M, Mizuno Y (1996) Cortical ballooned neurons in progressive supranuclear palsy. Neurosci Lett 209: 109-112
44. Martinez-Lage P, Munoz DG (1997) Prevalence and disease association of argyrophilic grains of Braak. J Neuropathol Exp Neurol 56: 157-164.
45. Tolnay M., Probst A (1998) Ballooned neurons expressing αB-crystallin as a constant feature of the amygdala in argyrophilic grain disease. Neurosci Lett 246: 165-168.
46. Lowe J, Errington DR, Lennox G et al. (1992) Ballooned neurons in several neurodegenerative diseases and stroke contain αB-crystallin. Neuropath Appl Neurobiol 18: 341-356.
47. Yamada T, McGeer PL, McGeer EG (1992) Some immunohistochemical features of argyrophilic grain dementia with normal cortical choline acetyltransferase levels but extensive subcortical pathology and markedly reduced dopamine. J Geriatr Psychiatry Neurol 5: 3-13.
48. De Vos RAI, Jansen ENH, Wesseling P (1996) Parkinson's disease and dementia with argyrophilic grains. Neuropathol Appl Neurobiol 22 (Suppl 1): 74-75.
49. Jellinger K, Bancher C (1998) Senile dementia with tangles (tangle predominant form of senile dementia). Brain Pathol 8: 367-376.
50. Ulrich J, Spillantini MG, Goedert M et al. (1992) Abundant neurofibrillary tangles without senile plaques in a subset of patients with senile dementia. Neurodegeneration 1: 257-284.
51. Kawashima T, Katsumi D, Iwaki T (1999) Argyrophilic grains in late-onset Creutzfeldt-Jakob diseased brain. Pathol Internat 49: 369-373.
52. Bancher C, Jellinger KA (1994) Neurofibrillary tangle predominant form of senile dementia of Alzheimer type: a rare subtype in very old subjects. Acta Neuropathol 88: 565-570.
53. Ikeda K, Akiyama H, Arai T et al. (1997) A subset of senile dementia with high incidence of the apolipoprotein E e2 allele. Ann Neurol 41: 693-695.
54. Bancher C, Egensperger R, Kösel S et al. (1997) Low prevalence of apolipoprotein E ε4 allele in the neurofibrillary tangle predominat form of senile dementia. Acta Neuropathol 94: 403-409.
55. Tolnay M, Probst A, Monsch AU (1998) Apolipoprotein E allele frequencies in argyrophilic grain disease. Acta Neuropathol 96: 225-227.
56. Ghebremedhin E, Schultz C, Botez G et al. (1998) Argyrophilic grain disease is associated with apolipoprotein E ε2 allele. Acta Neuropathol 96: 222-224.
57. Davis DG, Ross GW, Petrovitch H et al. (1997) Quantitation of argyrophilic grains in hippocampal CA1 of aged Japanese-American men. J Neuropathol Exp Neurol 56: 587.
58. Ulrich J, Johannson-Locher G, Seiler WO et al. (1997) Does smoking protect from Alzheimer's disease? Alzheimer-type changes in 301 unselected brains from patients with known smoking history. Acta Neuropathol 94: 450-454.
59. Tolnay M, Schwietert M, Monsch AU et al. (1997) Argyrophilic grain disease: distribution of grains in patients with and without dementia. Acta Neuropathol 94: 353-358.
60. Strittmatter WJ, Roses AD (1996) Apolipoprotein E and Alzheimer's disease. Annu Rev Neurosci 19: 53-77.
61. Ronald and Nancy Reagan Research Institute of the Alzheimer's Association and the National Institute on Aging Working Group (1998) Consensus report of the Working Group on "Molecular and Biochemical Markers of Alzheimer's Disease". Neurobiol Aging 9: 109-116.

62. Schneider JA, Gearing M, Robbins RS et al. (1995) Apolipoprotein E genotype in diverse neurodegenerative disorders. Ann Neurol 38: 131-135.
63. Botez G, Schultz C, Ghebremedhin E et al. (2000) Klinische Aspekte der "argyrophilic grain disease". Nervenarzt 71: 38-43.
64. Juottonen K, Laakso MP, Partanen K et al. (1999) Comparative MR analysis of the entorhinal cortex and hippocampus in diagnosing Alzheimer disease. Am J Neuroradiol 20: 139-144.
65. Frisoni GB, Laakso MP, Beltramello A et al. (1999) Hippocampal and entorhinal cortex atrophy in frontotemporal dementia and Alzheimer's disease. Neurology 52: 91-100.

TAU PATHOLOGY IN NEURONS AND GLIAL CELLS OF AGED BABOONS

[1]Christian Schultz, [2]Gene B. Hubbard, [1]Kelly Del Tredici, [1]Eva Braak, and [1]Heiko Braak
[1]Department of Clinical Neuroanatomy, Johann-Wolfgang Goethe-University, 60590 Frankfurt/Main, Germany, [2]Department of Laboratory Animal Medicine, Southwest Foundation for Biomedical Research, San Antonio, Texas U.S.A.

1. INTRODUCTION

The formation of cytoskeletal alterations associated with abnormally phosphorylated microtubule-associated tau protein is emerging as a common pathomechanism in several neurodegenerative diseases of the human brain. These disorders include Alzheimer's disease (AD), the most common and most extensively studied disorder associated with abnormal tau protein[1]. Tau pathology is also seen in other "tauopathies" including corticobasal degeneration, Pick's disease, progressive supranuclear palsy, and argyrophilic grain disease[2,3,4]. These diseases are characterized by tau-positive inclusions deposited not only in nerve cells but also in glial cells[3,5]. Finally, tau pathology develops in some families afflicted with hereditary frontotemporal dementia and parkinsonism linked to chromosome 17[6,7]. Interestingly, FTDP-17 recently has been linked to a number of pathogenetic mutations of the tau gene[8]. At this point, the mechanisms of tau pathology in neurons and glial cells are not fully understood. Animal models replicating characteristics of human tauopathies could provide information not obtainable from studies of the human brain. In an earlier report, we demonstrated the existence of neuronal and glial tau pathology in two out of four aged baboons[9]. Subsequently, we extended our sample to a total of 50

Neuropathology and Genetics of Dementia, Edited by Tolnay and Probst
Kluwer Academic/Plenum Publishers, New York 2001

brains of baboons ranging in age from 1 to 30 years. This allowed us to examine the effects of age on the development of tau pathology in baboons.

2. MATERIALS AND METHODS

A total of 50 brains of baboons (20 males, 30 females), aged 1-30 years, was examined. The majority of the brains (n=46) was obtained from baboons belonging to the subgenus *Papio anubis*. These baboon cases all were spontaneously generated in the 3600-member breeding colony at the Southwest Foundation for Biomedical Research (SFBR) in San Antonio, Texas. Baboons were group-housed, fed a commercial monkey diet, and provided with water *ad libitum*. All of the baboons in the colony were cared for in compliance with applicable animal care guidelines. The animals were born within the colony or came to the colony at less than five years of age. The brain tissue of all of these animals was obtained within one hour of euthanasia and generally within 20 minutes. The baboons were euthanized using intravenous commercial euthanasia agents. The animals had not been examined in any previous behavioral or experimental studies. Four brains included in a previous study were obtained from animals belonging to the subgenus *Papio hamadryas*[9]. These animals were raised and housed in zoological gardens in Germany. Their ages were estimated by history, dentition, and general appearance at the time of acquisition.

The brains were removed, bisected midsagitally, and fixed by immersion in formaldehyde for several weeks. One hemisphere was cut in the frontal plane into 1-2 cm thick sections. Serial coronal hemisphere sections (100 µm) containing the anterior hippocampal formation and the entorhinal cortex were cut using a freezing macrotome. Additional hemisphere sections were cut through the orbitofrontal cortex, central portions of the amygdala, lateral geniculate nucleus, and the occipital cortex.

Argyrophilic pathology was visualized using the modified Gallyas silver-iodide technique for neurofibrillary changes of the Alzheimer type[10,11]. Abnormally phosphorylated tau protein was detected by the monoclonal antibody AT8 (1:1000, Innogenetics, Ghent, Belgium) directed against the phosphorylated tau epitopes serine 202 and threonine 205[1,12,13]. Additional sections of each brain were stained using the phosphorylation-dependent antibody PHF-1[14] (1:250; gift of P. Davies) and the conformation-sensitive and phosphorylation-dependent antibody TG-3[15] (1:10; gift of P. Davies). Immunostaining was performed using the ABC-complex (Vectastain, Vector Laboratories, Burlingame, CA) method and 3, 3′-diaminobenzidine-tetrahydrochloride H_2O_2 (DAB, D5637 Sigma) as a chromogene. The severity of the cytoskeletal changes was rated using an AT8-immunostained hemisphere section cut at the rostrocaudal level of the mamillary body. The 50 baboon brains were categorized into three age groups: group I: 1-10 years (*n*=9), group II: 11-20 years (*n*=13), group III: 21-25 years (*n*=17), group IV: 26-30 years (*n*=11).

3. RESULTS

The cytoskeletal alterations in the baboons consist of neuronal, astroglial, and oligodendroglial inclusions[9]. The antibody AT8 labels the structures with sufficient detail to make possible their morphological classification. A substantial interindividual variability was noted in the overall density of the AT8-immunoreactive (AT8-ir) changes. This variability enabled us to categorize the density of the changes using four semiquantitative stages (Table 1). Mild changes were classified as 1-20 cellular structures containing abnormal tau protein. A severe involvement was detectable even with the unaided eye owing to the dense accumulation of AT8-ir structures (Figure 1). Two intermediate stages, namely moderate and marked affection, were introduced to categorize the transition from mild to severe involvement.

Age group	mild	moderate	marked	severe	absent
1-10 yrs *(n=9)*	0%	0%	0%	0%	100%
11-20 yrs *(n=13)*	31%	0%	0%	0%	69%
21-25 yrs *(n=17)*	35%	12%	12%	12%	29%
26-30 yrs *(n=11)*	36%	0%	18%	36%	9%

Table 1: Rating of AT8-ir changes in the 50 baboons. The categorization into age groups is described in the Materials and methods section. The cytoskeletal changes were classified using a semiquantitave scale: absent, mild, moderate, marked, or severe. Rating was performed for the overall amount of AT8-ir structures detectable in a 100 μm hemisphere section.

The distribution of the cytoskeletal alterations resembled that documented in our previous study[9]. In particular, the preferential involvement of allocortical medial temporal lobe structures was confirmed.

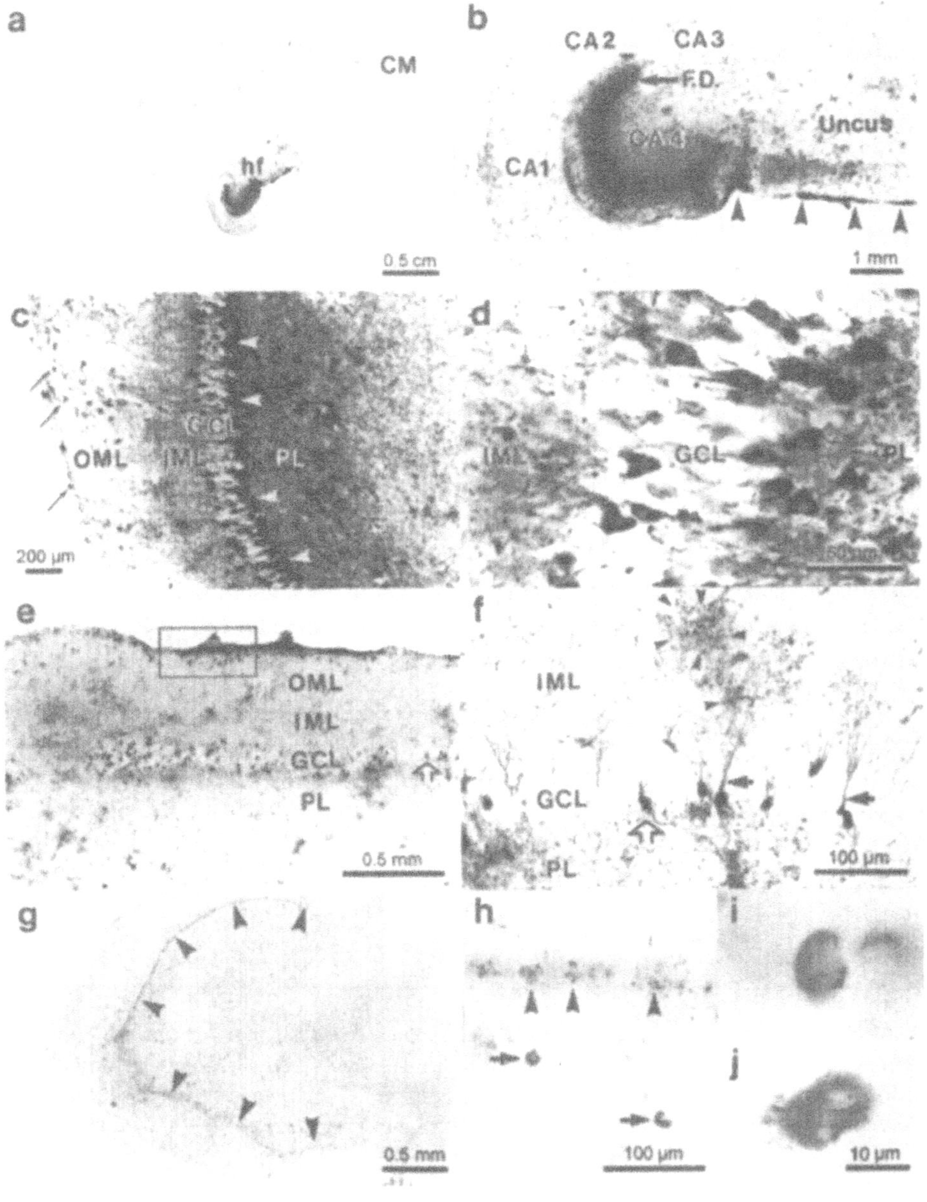

Figure 1. Cytoskeletal abnormalities in a 30-year-old baboon. AT8 immunostaining (a-f), Gallyas silver staining (**g-j**). **a:** Dense accumulation of AT8-positive cytoskeletal changes in the hippocampal formation (hf). **b:** A moderate density of AT8-positive inclusions is seen in the first and third sector of the Ammon's horn (CA1 and CA3). Cytoskeletal changes are even more pronounced in CA4,whereas CA2 is almost devoid of changes. The highest density of changes is noted in the fascia dentata (F.D.; arrow). Arrowheads point to subpial astroglial cytoskeletal changes bordering the hippocampal fissure (see also Fig. 4). **c:** Layer-specific pathology in the fascia dentata. AT8-positive inclusions in the outer molecular layer resemble

oligodendroglial coiled bodies. The inner molecular layer (IML) and the subgranular plexiform layer (PL) of the fascia dentata contain altered neuropil structures. Note the dense accumulation of altered neurons in the granule cell layer (GCL) which is particularly pronounced in the deeper granule cell layers (arrowheads). **d:** Cytoskeletal changes of the fascia dentata at higher magnification. The granule cell layer (GCL) contains numerous cellular inclusions bordered by the neuropil changes of the inner molecular layer (IML) and the plexiform layer (PL). **e:** Moderate degree of cytoskeletal changes in medial portion of the uncus. Intracellular inclusions occur predominantly in the granule cell layer (GCL). Open arrow points to an altered granule cell which is also denoted by an open arrow in the close-up view in (**f**). Framed box denotes a supial accumulation of astroglial changes. **f:** Cell bodies and dendrites of individual granule cells contain abnormal tau protein. Black arrows point to the apical dendritic stem of two diffusely labelled granule cells. A plaque-like accumulation of altered neuropil structures (arrowheads) accumulates in the inner molecular layer. The neuron labelled by the open arrow is also arrowed in (**e**). **g:** Layer-specific accumulation of Gallyas-positive neurofibrillary tangles (NFTs) in the granule cell layer (traced out by arrowheads). **h:** Accumulation of Gallyas positive-material in the granule cell layer (arrowheads). Single NFTs are seen in the hilus/CA4 region (arrows). **i:** Typical crescent-shaped NFT located in the granule cell layer. **j:** Large NFT in the hilus (arrow). Reproduced with permission from the Journal of Neuropathology and Experimental Neurology.

Cytoskeletal lesions were demonstrable in the granule cells of the fascia dentata (Figure 1). Neuronal AT8-ir changes also were detected in pyramidal neurons of the hippocampus proper with a slight preponderance in the first Ammon's horn sector (Figure 2a) as well as in projection neurons of the amygdala (Figure 2b). In four animals a moderate or marked accumulation of AT8-ir material was noted to preferentially affect the superficial entorhinal cell layer II (Pre-ß). These changes did not extend into the subiculum and sharply decreased in density at the lateral border of the entorhinal territory marked by the rhinal sulcus. Astroglial changes showed a less consistent distributional pattern but generally were most pronounced in the CA4 region and in the subgranular plexiform layer beneath the granule cell layer of the fascia dentata. Astroglial involvement also was found in subependymal regions of the lateral ventricle located around the inferior horn as well in tissues surrounding the third ventricle (Figure 3). Subpial astroglial changes were noted along the periamygdaloid cortex and the basal mamillary region. Oligodendroglial inclusions, referred to as coiled bodies, were distributed chiefly throughout the termination zone of the entorhinal perforant path, the fimbria hippocampi, and the fornix (Figure 4).

As detailed below, the occurrence of glial changes was noted even in baboons with only mild cytoskeletal changes. In all of the baboons with either marked or severe cytoskeletal pathology, the changes consisted of a combination of astroglial, oligodendroglial, and neuronal inclusions. The prevalence of the different stages of overall AT8-ir tau pathology assessed for each of the four age categories (Table 1). All nine brains in group I (1-10 years) were devoid completely of AT8-ir structures. In the brains of group II, animals 11-20 years, cytoskeletal alterations were detected in four cases (31%). The changes in these animals were rated as mild and consisted of small clusters of AT8-ir astrocytes in three instances or of a single AT8-ir neuron in the entorhinal cortex of one animal. In group III (21-25 years), 12 animals (71%) exhibited some amount of AT8-ir changes. In two animals belonging to this group (aged 21 years and 23 years) the changes were rated as severe. In group IV (26-30 years), 10 out of 11 (91%) brains exhibited the presence cytoskeletal alterations. The lesions were rated as severe in three of them (36%).

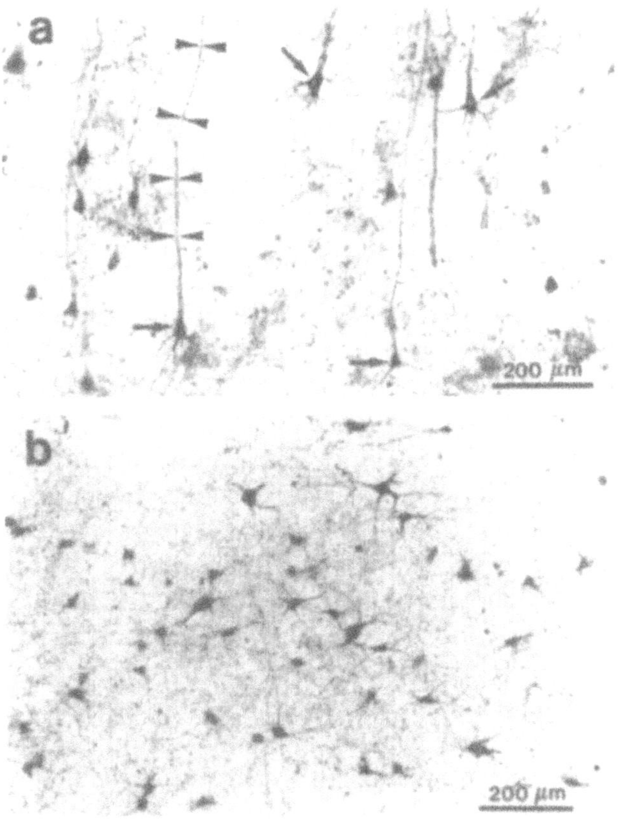

Figure 2. Cytoskeletal pathology in neurons of the baboon labelled by AT8-immunostaining **a:** Deposition of abnormal tau in pyramidal neurons of the first Ammon's horn sector (CA1) Abnormal tau is deposited in cell bodies (arrows) and apical dendrites (arrowheads) of CA1 pyramidal neurons (100 µm, 30-year-old baboon). **b:** Diffusely labelled neurons of the amygdala (100 µm, 26-year-old baboon). Reproduced with permission from the Journal of Neuropathology and Experimental Neurology.

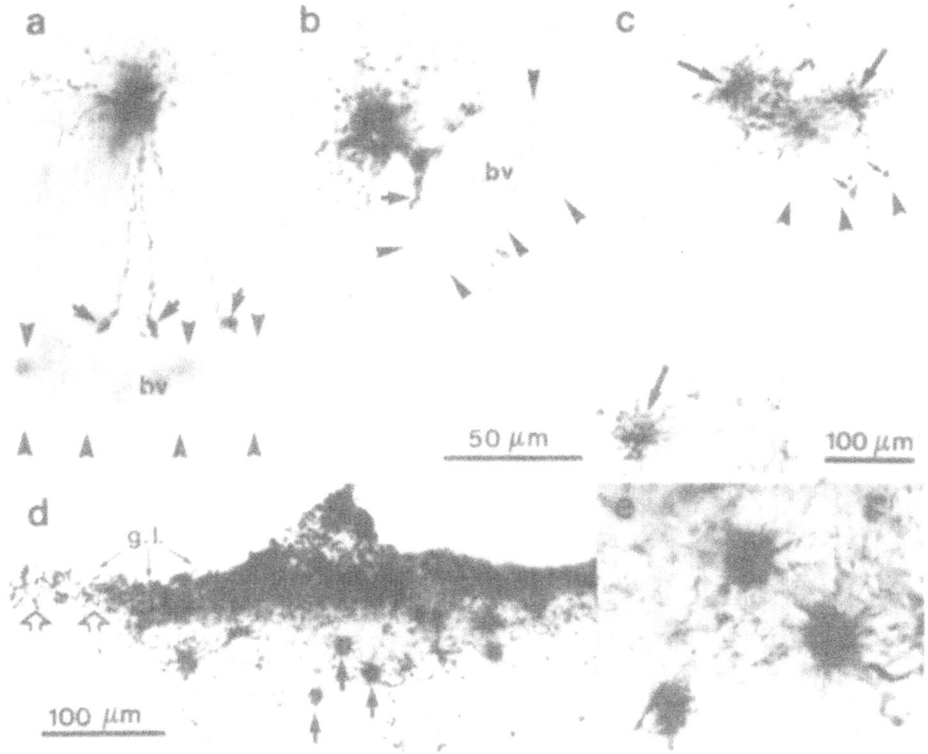

Figure 3. Astroglial pathology in a 30-year-old baboon. Abormally phosphorylated tau is demonstrated in putative astrocytes using AT8 immunostaining. **a-c:** Abnormal tau protein in perivascular astrocytes. The astrocyte in (**a**) gives rise to three processes abutting with end feet (arrows) upon the wall of a blood vessel (bv, arrowheads). **b:** A perivascular astroglial process (arrow) contacts the wall of a blood vessel (bv, arrowheads). Scale bar in (**b**) is also valid for (**a**). **c:** Abnormal tau protein is diffusely deposited in three astrocytes (arrows). One of them (upper right corner) gives rise to thin branching processes terminating with end feet upon a small blood vessel (arrowheads) **d:** Subpial tau-positive astrocytes (thick arrows, close-up view in **e**) accompanied by a dense network of altered processes in the glial limiting membrane (g. l., thin arrows). Open arrows point to a segment of the glial limiting with only mild cytoskeletal changes not accompanied by the subpial tau-positive astrocytes. Reproduced with permission from the Journal of Neuropathology and Experimental Neurology.

4. DISCUSSION

This study demonstrates an age-related progression of tau pathology in brains of baboons. The rationale for this study comes from our prior documentation of abnormal tau protein in neurons and glial cells of two aged baboons[9]. The previous sample was limited by the small number of animals available to us (n=4). Here, we present data from 50 baboons aged 1 to 30 years.

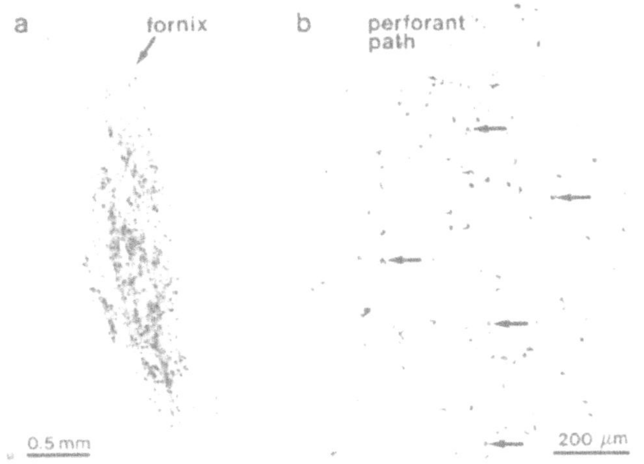

Figure 4. Cytoskeletal changes in the white matter immunostained by AT8. **a:** 26-year-old baboon; **b:** 30-year-old baboon. **a:** Abnormal tau in the fornix (hypothalamus). The changes consist of interfascicular fibers and cellular inclusions corresponding to oligodendroglial coiled bodies. **b:** Oligodendroglial coiled bodies (arrows) in the perforant path of a 30-year-old baboon. Reproduced with permission from the Journal of Neuropathology and Experimental Neurology.

We demonstrated that AT8-positive tau pathology is absent in young baboons aged 1 to 10 years. None of the nine animals included in this age group exhibited any discernible AT8-positive structures. This finding clearly contradicts the view that the AT8-positive structures in baboons result from cross-reactions with normal tau epitopes, which should be detectable in baboons of all ages.

Second, we determined that the onset age of tau pathology in baboons is approximately 19 years. Three out of five animals at age 19 years exhibited mild to moderate cytoskeletal alterations, whereas all animals below that age were devoid of changes. The existence of even mild cytoskeletal changes should not be neglected, because these alterations provide insights into the evolution of tau pathology in baboons. For example, glial pathology was noted in the majority of animals with only mild or moderate lesions. Such early glial pathology is suggestive of pathogenetic pathways targeting not only the neuron but also primarily affecting glial cells. Elucidation of these pathways may help to understand the role of glial tau pathology developing in human tauopathies[2,3,5,16,17]. Third, progression of tau pathology in baboons gathers considerable momentum above the age of 20 years. Eleven out of 17 animals (65%) between 21-25 years of age exhibited some degree of tau pathology. In addition to mild and moderate changes, this age group also includes cases with marked or severe involvement. Thus, the youngest animal in our present sample showing severe changes was 21 years old (Table 1).

Finally, the oldest baboons available to us (26-30 years) show a further increase in prevalence and severity of the cytoskeletal alterations. The presence of tau pathology was noted in 11 out of 12 animals (91%), with only one baboon remaining completely devoid of lesions. Moreover, this age group also includes the majority of animals with lesions rated as severe (n=4 [36%]). Accordingly, the presence of pronounced tau pathology in baboons is not a rare finding. Instead, it can be encountered in approximately every third animal aged 26 years and older.

In agreement with the findings discussed above, our statistical analysis shows a significant correlation between advancing age and increasing severity of tau pathology in baboons. In conclusion, age is an important risk factor for the progression of tau pathology in baboons. Similarly, age is acknowledged as the single most important and universal risk factor for sporadic tauopathies affecting the human brain. In particular, advancing age strongly correlates with an increasing prevalence and severity of Alzheimer-related tau pathology[18]. Further examination of age-related tau pathology in baboons may help to pinpoint the age-related factors that confer on the human brain vulnerability to tau pathology.

In summary, baboons begin to manifest tau pathology at the end of their second decade of life and experience a marked progression of the cytoskeletal alterations during their third decade. This age-related emergence along with the high frequency of the lesions, the coexistence of glial and neuronal involvement, and the phenomenon of selective vulnerability constitute the features that render the baboon a potential primate model for the ongoing study of age-related human tauopathies. The ultimate value of this model depends on its further characterization at the ultrastructural, molecular biological, genetic, and behavioral levels.

ACKNOWLEDGEMENTS

This study was supported by the Deutsche Forschungsgemeinschaft, the Bundesministerium für Bildung, Wissenschaft, Forschung und Technologie. The generous gift of the antibodies PHF-1 and TG3 provided by Dr. P. Davies (Department of Pathology, Albert Einstein College, NY) is gratefully acknowledged. The technical support of A. Biczysko, M. Babl, A. Börner, U. Fertig, S. McAnn, M. Silva, I. Szasz (graphics) is gratefully acknowledged.

REFERENCES

1. Braak E, Braak H, Mandelkow EM (1994) A sequence of cytoskeleton changes related to the formation of neurofibrillary tangles and neuropil threads. Acta Neuropathol 87: 554-567.
2. Braak H, Braak E (1998) Argyrophilic grain disease: frequency of occurrence in different age categories and neuropathological diagnostic criteria. J Neural Transm 105: 801-809.
3. Feany M.B, Mattiace, LA, Dickson DW (1996) Neuropathologic overlap of progressive supranuclear palsy, Pick´s disease and corticobasal degeneration. J Neuropathol Exp Neurol 55: 53-67.
4. Tolnay M, Spillantini MG, Goedert M et al. (1997) Argyrophilic grain disease: widespread hyperphosphorylation of tau protein in limbic neurons. Acta Neuropathol 93: 477-484.
5. Chin SSM, Goldman J (1996) Glial inclusions in CNS degenerative diseases. J Neuropathol Exp Neurol 55: 499-508.
6. Foster NL, Wilhelmsen K, Sima AA et al. (1997) Frontotemporal dementia and parkinsonism linked to chromosome 17: a consensus conference. Ann Neurol 41: 706-715.
7. Spillantini M.G, Bird TD, Ghetti B (1998) Frontotemporal dementia and Parkinsonism linked to chromosome 17: a new group of tauopathies. Brain Pathol 8: 387-402.
8. Hutton M, Lendon CL, Rizzu P et al. (1998) Association of missense and 5'-splice-site mutations in tau with the inherited dementia FTDP-17. Nature 393: 702-705.
9. Schultz C, Dehghani F, Hubbard GB et al (2000) Filamentous tau pathology in nerve cells, astrocytes and oligodendrocytes of aged baboons. J Neuropathol Exp Neurol 59: 39–52.
10. Braak H, Braak E (1991) Demonstration of amyloid deposits and neurofibrillary changes in whole brain sections. Brain Pathol 1: 213-216.
11. Gallyas F (1971) Silver staining of Alzheimer's neurofibrillary changes by means of physical development. Acta Morphol Acad Sci Hung 19: 1-8.
12. Mercken M, Vandermeeren M, Lübke U et al. (1992) Monoclonal antibodies with selective specificity for Alzheimer tau are directed against phosphatase-sensitive epitopes. Acta Neuropathol 84: 265-272.
13. Goedert M, Jakes R, Vanmechelen E (1995) Monoclonal antibody AT8 recognizes tau protein phosphorylated at both serine 202 and threonine 205. Neurosci Lett 189: 167-170.
14. Greenberg SG, Davies P, Schein JD et al. (1992) Hydrofluoric acid-treated tau/PHF proteins display the same biochemical properties as normal tau. J Biol Chem 267: 564-569.
15. Jicha GA, Lane E, Vincent I et al. (1997) A conformation- and phosphorylation-dependent antibody recognizing the paired helical filaments of Alzheimer's disease. J Neurochem 69: 2087-2095.
16. Schultz C, Koppers D, Sassin I et al. (1998) Cytoskeletal alterations in the human tuberal hypothalamus related to argyrophilic grain disease. Acta Neuropathol 96: 596-602.
17. Botez G, Probst A, Ipsen S et al. (1999) Astrocytes expressing hyperphosphorylated tau protein without glial fibrillary tangles in argyrophilic grain disease. Acta Neuropathol 98: 251-256.

18. Braak H, Braak E (1997) Frequency of stages of Alzheimer-related lesions in different age categories. Neurobiol Aging 18: 351-357.

HUMAN TAU TRANSGENIC MICE

Towards an animal model for neuro- and glialfibrillary lesion formation

[1,*]Jürgen Götz, [2]Markus Tolnay, [1]Robi Barmettler, [1]Alessandra Ferrari, [3]Kurt Bürki, [4]Michel Goedert, [2]Alphonse Probst, and [1]Roger M. Nitsch

*[1] Division of Psychiatry Research, University of Zurich, August Forel Strasse 1, 8008 Zurich, Switzerland; [2]Institute of Neuropathology, University of Basel, Schönbeinstrasse 40, 4003 Basel, Switzerland; [3]Institute of Laboratory Animal Science, University of Zürich, Winterthurer Strasse 190, Switzerland; [4]MRC Laboratory of Molecular Biology, Hills Road, Cambridge CB2 2QH, UK; *e-mail: goetz@bli.unizh.ch; http://www.neuroscience.unizh.ch/e/groups/goetz00.htm*

1. INTRODUCTION

Abundant neurofibrillary and glial fibrillary lesions composed of hyperphosphorylated microtubule-associated tau protein constitute the major cellular pathology of several neurodegenerative diseases, the so-called tauopathies. These include Alzheimer's disease, Pick's disease, progressive supranuclear palsy, corticobasal degeneration and frontotemporal dementia with Parkinsonism linked to chromosome 17 (FTDP-17)[1].

Tau is an abundant protein in both central and peripheral nervous systems. In brain it is found predominantly in neurons, where it is concentrated in axons. Tau has microtubule-binding and tubulin-polymerizing activities *in vitro*, and it establishes short crossbridges between axonal microtubules. In addition, it controls intracellular trafficking[2]. In adult human brain, six tau isoforms are produced from a single gene by alternative mRNA splicing. They differ by the presence or absence of one or two short inserts in the amino-terminal half, and have either three or four microtubule-binding repeats in the carboxy-terminal half. In contrast to

Neuropathology and Genetics of Dementia, Edited by Tolnay and Probst
Kluwer Academic/Plenum Publishers, New York 2001

humans, mice express only four-repeat isoforms[3]. Tau is a phosphoprotein and phosphorylation is developmentally regulated in that tau from immature brain is phosphorylated at more sites than tau from adult brain.

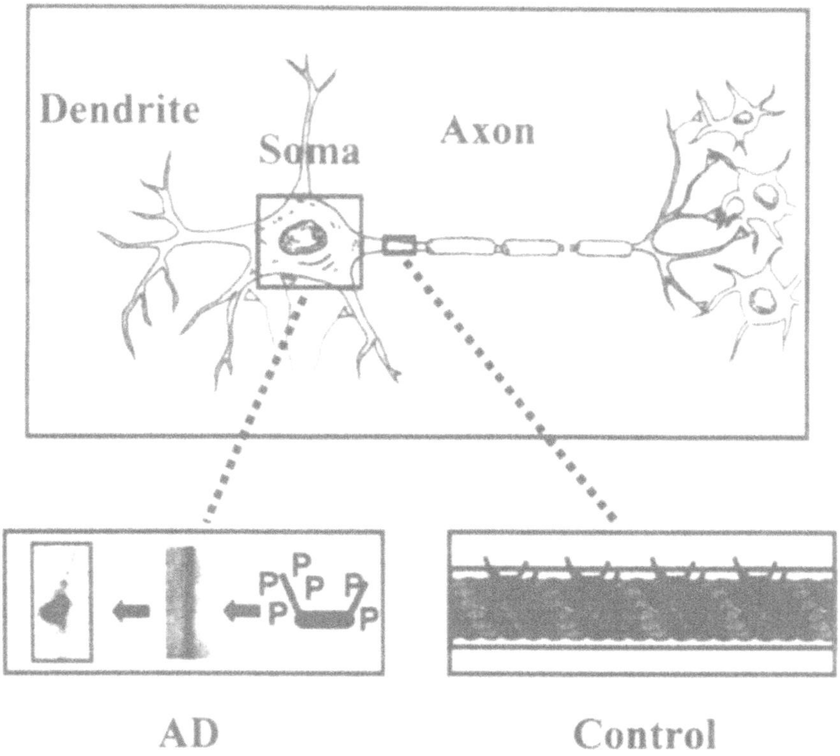

Figure 1. Under physiological conditions, tau is localized to the axon where it binds to microtubules and polymerizes tubulin by establishing short crossbridges between axonal microtubules. Under pathological conditions such as AD, tau is hyperphosphorylated (P) and relocalized from axonal to somatodendritic compartments where it assembles into filamentous aggregates. In advanced stages of disease, these filaments fill up the entire neuronal soma.

In Alzheimer's disease and other tauopathies, brain tau is hyperphosphorylated and is found not only in axons, but also in cell bodies and dendrites of affected nerve cells[4] (Figure 1). In contrast to Alzheimer's disease, where filaments are found only in neurons, numerous tau filament-containing glial cells are present in a variety of tauopathies including progressive supranuclear palsy, corticobasal degeneration and argyrophilic grain disease. At present, it is unknown, whether the glial pathology affects degeneration of neurons and whether it is required for progression of disease[5,6].

Alzheimer's disease is characterized by a good correlation between the presence of tau pathology and the degree of cognitive impairment, indicating a close link between the development of tau pathology and nerve cell degeneration. The formation of neurofibrillary lesions may pre-date amyloid plaque formation, a histopathological hallmark of Alzheimer's disease.

The recent discovery of mutations in exonic and intronic regions in the tau gene in FTDP-17 has established that dysfunction of tau can cause neurodegeneration, and lead to dementia[7]. Mutations in the tau gene in FTDP-17 result either in the reduced ability of tau to interact with microtubules, or to a change of the normally existing ratio of four-repeat (4R) to three-repeat (3R) tau, resulting in the net overproduction of 4R tau isoforms. These observations suggest that either mere overproduction of 4R tau or the presence of missense mutations (point mutations and deletions) are sufficient to cause frontotemporal dementia.

Currently, it is unknown whether the presence of tau filaments causes degeneration of neurons in tauopathies or whether filament formation is only a consequence of a more general degenerative process. A causative role for filaments in neurodegeneration is not unlikely: For example, in $\alpha 1$-antitrypsin-associated liver disease hepatocyte loss and cirrhosis is clearly a consequence of abnormal deposits and not of a loss of function, because cirrhosis develops only with a conformationally unstable variant, but not with mutations that cause "null" suppressions of synthesis[8]. Mice lacking tau appear to be immunohistologically normal, suggesting that other microtubule-binding proteins including MAP-1A can functionally compensate for the loss as shown by intact axonal elongation in cultured neurons. But in some small-calibre axons, microtubule stability is decreased, and microtubule organization is significantly changed[9]. Interestingly, mice lacking tau show a phenotype that parallels some signs and symptoms of FTDP-17 patients. They develop muscle weakness in the wire-hanging test, hyperactivity in a novel environment, and impairment in the contextual fear conditioning. They also had a tendency to fall more easily in the rod-walking test. This suggests that the loss of tau protein may itself lead to some of the neurological characteristics observed in FTDP-17 patients[10].

Therefore, in order to reproduce key aspects of tau pathology in transgenic mice, we chose three alternative approaches: (1) overexpression of wild-type 4R human tau in neuronal cells, (2) overexpression of mutant human tau in neuronal and glial cells, and (3) overexpression of dominant negative mutant forms of protein phosphatase 2A, a phosphatase known to dephosphorylate tau *in vitro* at epitopes which are diagnostic for AD.

2. TRANSGENIC AND KNOCKOUT APPROACHES

2.1 Hyperphosphorylation and somatodendritic localization of tau in wild-type human tau transgenic mice (ALZ 7 line)

Expression of the longest human four-repeat (4R) tau isoform under the control of the murine Thy1 promoter along with a polyadenylation/splice cassette derived from the β-globin gene (ALZ 7 construct) resulted in the presence of hyperphosphorylated tau not only in axons, but also in nerve cell bodies and dendrites. Furthermore, tau was phosphorylated at sites that are known to be phosphorylated in paired helical filaments, the principal fibrous component of neurofibrillary lesions including neurofibrillary tangles, neuropil threads and dystrophic neurites[11]. A similar phenotype was also reported in mice with neuronal expression of the shortest human tau isoform[12]. In summary, these mouse strains showed early changes associated with the development of neurofibrillary lesions in Alzheimer's disease and related disorders, but they failed to produce neurofibrillary tangles and lacked neurological symptoms.

2.2 Neurological symptoms in wild-type human tau transgenic mice (ALZ 17 line)

Because the amyloid plaque load in transgenic mice expressing FAD mutations of human amyloid precursor protein APP was directly correlated with the expression level of the transgene[13], we chose a stronger murine Thy1.2 expression vector for tau expression. Two ALZ 17 lines with high expression levels of the longest human 4R tau isoform were established[14]. Tau mRNA levels were approximately fivefold higher than in the previously described ALZ 7 line[11]. Tau-positive nerve cells were observed in brain stem, spinal cord, cerebral cortex and the hippocampus, including CA1, dentate gyrus, as well as in the entorhinal cortex. We found in brains of these mice transgenic tau that reacted with a number of phosphorylation-dependent anti-tau antibodies. In particular, we found both a highly compartmentalized phosphorylation of tau at specific epitopes, and a pattern of tau phosphorylation which varied among distinct neuronal cell types[15]. Gallyas silver impregnations[16], however, failed to identify classical neurofibrillary lesions. To determine the solubility of tau, spinal cords from up to 12 month-old mice were extracted in high-salt RAB buffer, RIPA buffer, and formic acid buffers to allow for the extraction of increasingly insoluble proteins. Unlike tau from control mice, a substantial amount of transgenic tau was also present in the RIPA-soluble and in the formic acid-soluble fractions. These findings indicate a markedly reduced solubility of tau from ALZ 17 mice relative to tau from control animals. This is, again, similar to pathological tau extractable from brains of patients with AD or other tauopathies.

Tau-immunoreactive neurons and axonal spheroids were already present in three week-old mice, and their number slightly increased with age. When brain sections were stained with anti-neurofilament antibodies, multiple spheroids with 7-10 μm in diameter were evident. Spheroids consisted of focal dilatations of axons and were found in neocortical areas with many neurons, which were stained with the phosphorylation-dependent antibody AT8. The spinal grey matter from homozygous mice of the ALZ 17 line contained large numbers of nerve cell bodies that were strongly immunostained by phosphorylation-dependent antibodies. In contrast to spheroids in the cerebral cortex, the majority of spinal cord spheroids was strongly immunoreactive with a number of anti-tau antibodies. Axonal spheroids similar to those found in ALZ 17 mice are a neuropathological characteristic of the majority of cases of sporadic and familial amyotrophic lateral sclerosis (ALS), where they are believed to impair slow axonal transport[17].

Evidence for axonal degeneration, including axonal breakdown and segmentation of myelin into ellipsoids (so-called "digestion chambers") was obtained in anterior spinal roots. Only few AT8-positive axons were detected in interfascicular nerve bundles of skeletal muscle. However, clear evidence of neurogenic muscle atrophy, with groups of small angular muscle fibers, was present in the hindleg musculature of transgenic, but not control, mice.

Electron microscopy studies revealed that axonal spheroids were surrounded by a thin myelin sheath made of only few myelin leaflets. The axoplasm was filled with loosely arranged 10 μm thick filaments and microtubules, with only few mitochondria in between. In spinal cord, in addition to axonal spheroids, there was evidence of axonal degeneration with collapsed and fragmented myelin sheaths.

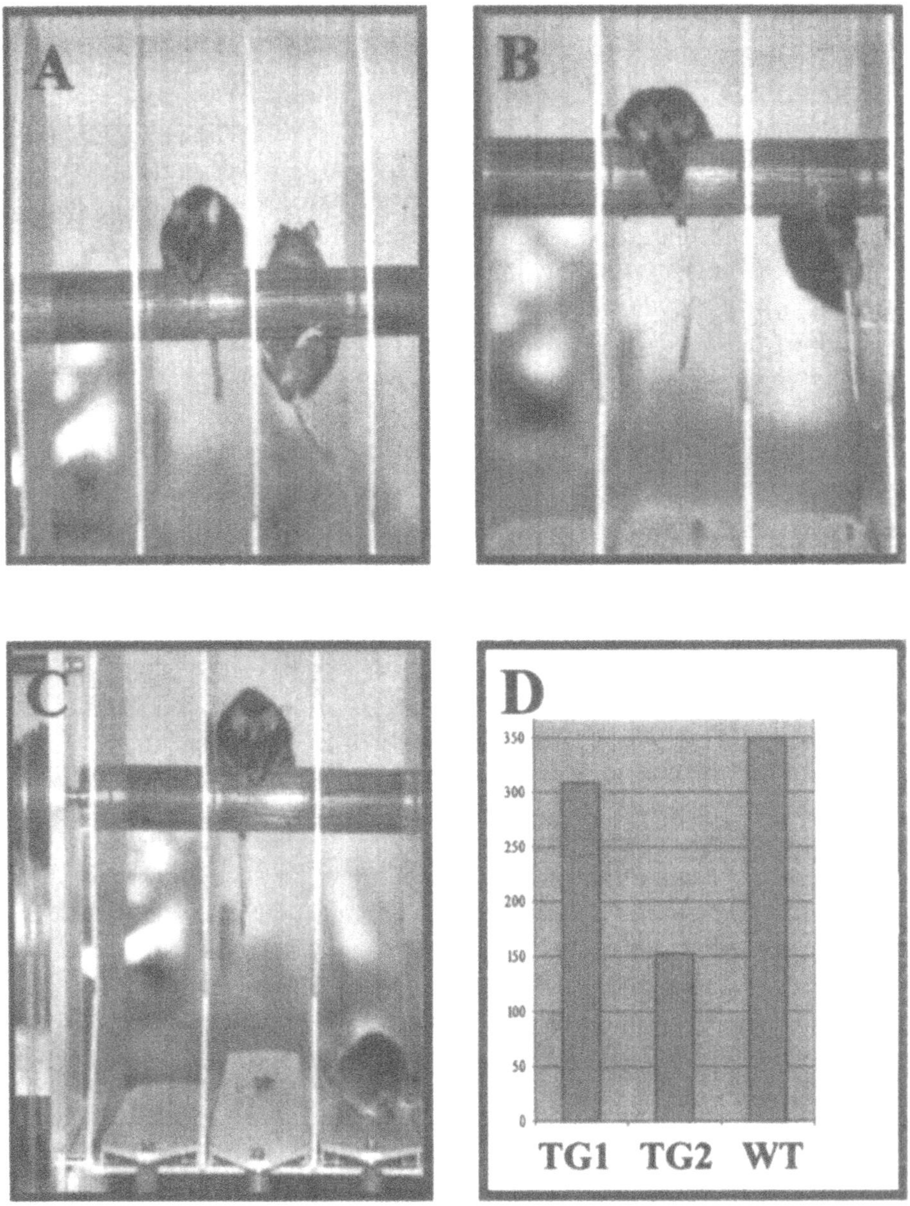

Figure 2. Abnormal motor coordination on the rotarod. In the acceleration mode, tau transgenic mice fall off the rotarod much earlier as compared to wild-type litter mates. With increasing age, the performance of transgenic mice worsens.

Transgenic mice appeared to be normal during the first few weeks of postnatal life. Then, progressively, they showed signs of neurological malfunction. When lifted by the tail, normal mice extend their hindlimbs and spread their toes. At one to three months of age, homozygous ALZ 17 mice extended their legs, but retracted their paws, in marked contrast to control mice. When aged 8-16 months, the same mice were unable to extend their hindlimbs when lifted by the tail. When tested on the rotarod, ALZ 17 mice showed abnormal motor coordination[14] (Figure 2).

Our data establish that overexpression of human 4R tau is sufficient to cause nerve cell dysfunction, but we detected neither neurofibrillary lesions nor obvious losses of neurons in these mice. These results were confirmed by other laboratories[18,19]. It appears likely that tau filament formation is necessary for massive neuronal cell loss as observed in FTDP-17.

2.3 Expression of FTDP-17 mutant human tau in transgenic mice

In order to reproduce neuro- and glialfibrillary tau lesion formation *in vivo*, we established a series of transgenic mice which express the FTDP-17 linked mutations P301L and G272V in neurons, oligodendrocytes, or astrocytes[20]. Mutations P301L and G272V tau were selected on the basis of *in vitro* experiments which had shown, compared to several other FTDP-17-associated tau mutants, both a higher tendency to aggregate into filaments and a lower affinity for tubulin.

2.3.1 Autoregulatory tetracycline-dependent transactivator loop system: Targeting of tau expression to neurons and oligodendrocytes

Tetracycline-regulated inducible gene expression relies on expression of a tetracycline-controlled transactivator (TA) fusion protein composed of the tetracycline repressor and the transcriptional activation domain of VP16. To achieve high expression levels of human tau in transgenic mice, we chose an autoregulatory tetracycline-dependent TA system as initially described by Shockett et al.[21] and generated two sets of transgenic lines: One expressed the tetracycline-dependent TA under control of a murine PrP promoter and directed its expression to both neuronal and glial cells; the second contained two TA-binding motifs (tetOp) in tandem, one upstream of a TA gene, the other upstream of a cDNA encoding G272V tau (Figure 3).

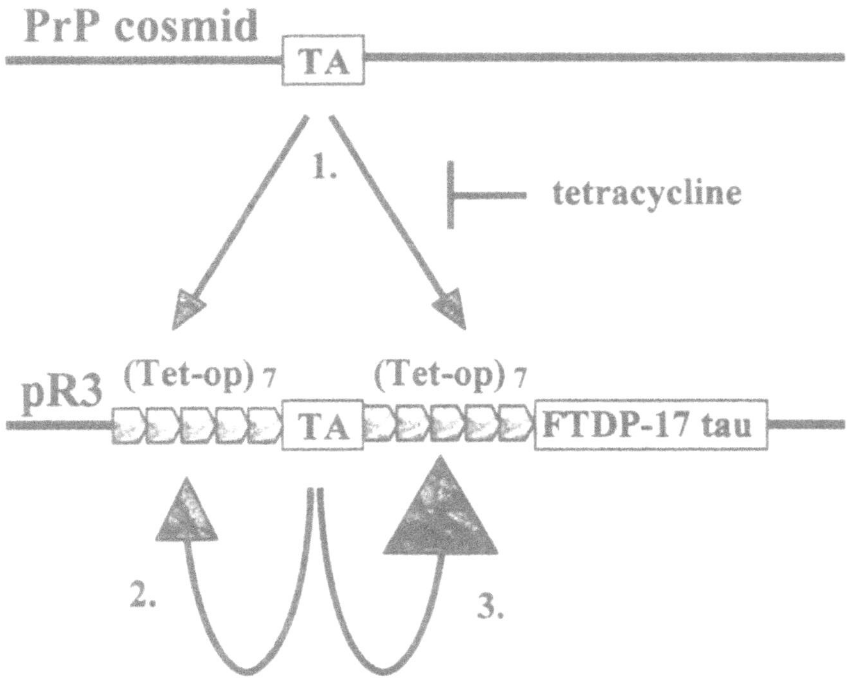

Figure 3. Autoregulatory loop system to direct high expression of FTDP-17 tau to those cells in which the PrP cosmid promoter is active. This approach is not only designed to achieve high expression levels, but also to suppress tau expression at any given time by the addition of tetracycline/doxycycline to the drinking water.

While single transgenic mice did not express G272V tau, double transgenic lines expressed high levels of G272V tau in a subset of neurons and glial cells in which the transgenic PrP promoter was active. In double transgenic mice, we found massive, clod-like, pre-tangle-type accumulations of tau both in neuronal cell bodies and dendrites and in glial cell bodies and processes. These were obvious already in three week-old mice. At six months of age, there was evidence for tau filament formation in some tau-expressing cells. Preliminary analyses of the functional consequences related to filament formation in these mice suggest selective induction of apoptosis as indicated by TUNEL-positive stainings (Götz et al., in preparation).

2.3.2 Targeting of tau expression to the nucleus

A second approach is based on the finding that incubation of recombinant tau with either sulfated glycosaminoglycans or RNA results in bulk assembly of tau into AD-like filaments[22]. Sulfated glycosaminoglycans and RNA share a repeat sugar backbone and negative charges in the form of sulfates or phosphates. Binding of heparin or RNA to tau may induce or stabilise a conformation of tau that brings the microtubule-binding repeats of individual tau molecules into close proximity, creating sites which favour polymerization into filaments.

Because of the highly compartmentalized concentration of RNA in nuclei, and because of the absence of its physiological binding partner, the microtubules, we designed two constructs (pR7 and pR8) that target expression of transgenic tau (P301L and G272V, respectively) to the nucleus. Despite the presence of a nuclear localization signal NLS, most of the tau in pR7 and pR8 transgenic lines was present in the cytoplasm. Highest levels of human tau were found in brain stem and spinal cord, where we expect to find neuronal filaments as the mice become older. It remains to be seen, whether nuclearly localized tau is more prone to filament formation than cytoplasmically localized tau. When tested on the rotarod (treadmill), pR8 mice showed, similar to the ALZ17 mice, abnormal motor coordination (Götz et al., in preparation). Therefore, pR8 mice will be analyzed by several additional paradigms including the narrow beam test, the grid walk test, and grip strength measurements.

2.3.3 Targeting of tau expression to astrocytes

The third approach is designed to understand the role of astrocytic tau filament formation in neurodegeneration. Expression of P301L and G272V tau was targeted to astrocytes by either using the GFAP or the PrP promoter. Levels of tau expression as determined by immunoblots were at least tenfold lower as compared to mice with neuronal expression of human tau, partly reflecting different promoter strengths. Human tau-positive astrocytes were found in several brain regions including cerebellum, and in spinal cord. Tau was phosphorylated at threonine 181 as shown with phosphorylation-dependent antibody AT270 (Ferrari et al., in preparation). Silver impregnation techniques and immunoelectron microscopy of sarcosyl-extracted tau will tell whether murine astrocytes are capable of forming filaments.

2.3.4 Targeting of tau expression to neurons

A forth approach uses the murine Thy1.2 expression vector to target P301L tau expression to neurons. This expression vector has also been used to generate the wild-type human tau (ALZ 17) transgenic mice. Transgenic lines were established, some of which resulted in highly restricted expression of P301L tau to the hippocampus. Tau was expressed in axonal and somatodendritic compartments of CA1 pyramidal and DG granule cells. In addition, the accumulated tau protein was phosphorylated at distinct epitopes (Figure 4). We expect that the accumulated tau will form filaments. Therefore, it will be important to determine by immunohistochemistry the appearance of the different phosphoepitopes as a function of age, cell type, and neuronal compartments (somatic, dendritic, axonal) and relate it to filament formation. Immunoelectron microscopy of sarcosyl-extracted brain tau will quantify the extent of phosphorylation at specific sites.

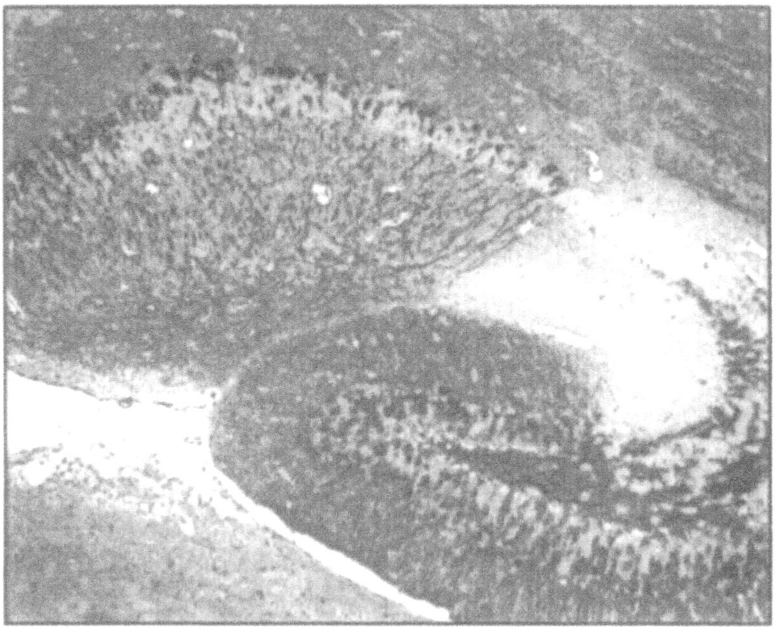

Figure 4. Phosphorylation-dependent antibody AD2 reveals tau expression in CA1 pyramidal neurons and dentate gyrus granule cells of P301L tau transgenic mice.

2.4 Expression of dominant negative mutant forms of PP2A in transgenic mice

Careful staging of brains has shown, that tau phosphorylation precedes filament formation[23]. *In vitro*, phosphorylated tau fails to assemble into AD-like tau filaments. On the other hand, sulfated glycosaminoglycans stimulate tau phosphorylation at lower concentrations than those required for filament formation[22]. Several kinases including MAPK, cdk5, cdc2, JNK, ERK2, MARK, PKA, GSK-3, CaMKII, and CKI, and phosphatases including PP1, PP2A, and PP2B have been suggested to play a role in tau phosphorylation *in vivo*. Among the phosphatases which dephosphorylate tau *in vitro*, protein phosphatase 2A (PP2A) has been shown to be associated with microtubules[24].

Protein phosphatase 2A holoenzymes are implicated in many different facets of cellular function including signal transduction pathways involved in apoptosis, growth control[25], neural development[26], and tau hyper-phosphorylation[4]. To address these functions *in vivo*, we generated a knockout of a PP2A subunit with crucial functional importance: PP2A catalytic subunit Cα knockout mouse embryos failed to develop mesoderm, and died around embryonic day 6.5[25]. During early development, Cα is localized at the plasma membrane, and assembles in a complex that contains E-cadherin and β-catenin[27]. Incidentally, β-catenin exists in a complex with presenilin 1, the gene of which is mutated in familial forms of AD. In order to study the role of PP2A in post-mitotic neurons in the adult nervous system, with particular emphasis on tau phosphorylation, we expressed dominant negative mutant forms of PP2A Cα under the control of a neuron-specific promoter and found that endogenous tau was phosphorylated at distinct sites[28]. These data support the view that PP2A function is essential in dephosphorylating tau, and that the phosphorylation status of each phosphoepitope of tau is regulated by specific phosphatases and kinases, as suggested by the phosphorylation pattern in human wild-type tau transgenic mice[15].

3. OUTLOOK

Our transgenic mice will be suitable to determine which phosphoepitopes of tau are important for filament formation *in vivo*, which cells are susceptible to filament formation, which subcellular compartments are capable of making tau filaments *in vivo*, how filaments interfere with neuronal functions including axonal transport, and whether glial cells with tau filaments can influence neuronal functions. They are also suited to determine whether other protein deposits including amyloid can accelerate tau pathology *in vivo*.

Eventually, mice with histopathological and behavioral changes will be available for drug testing aiming to halt or prevent AD and related tauopathies.

REFERENCES

1. Spillantini MG, van Swieten JC, Goedert, M (2000) Tau gene mutations in frontotemporal dementia and parkinsonism linked to chromosome 17 (FTDP-17). Neurogenetics 2: 193-205.
2. Trinczek B, Ebneth A, Mandelkow EM et al. (1999) Tau regulates the attachment/detachment but not the speed of motors in microtubule-dependent transport of single vesicles and organelles. J Cell Sci 112: 2355-2367.
3. Goedert M, Jakes R (1990) Expression of separate isoforms of human tau protein: correlation with the tau pattern in brain and effects on tubulin polymerization. EMBO J 13: 4225-4230.
4. Goedert M, Spillantini MG, Jakes R et al. (1995) Molecular dissection of the paired helical filament. Neurobiol Aging 16: 325-334.
5. Komori T (1999) Tau-positive glial inclusions in progressive supranuclear palsy, corticobasal degeneration and Pick's disease. Brain Pathol 9: 663-679.
6. Delisle MB, Murrell JR, Richardson R et al. (1999) A mutation at codon 279 (N279K) in exon 10 of the Tau gene causes a tauopathy with dementia and supranuclear palsy. Acta Neuropathol 98: 62-77.
7. Hutton M, Lendon CL, Rizzu P et al. (1998) Association of missense and 5'-splice-site mutations in tau with the inherited dementia FTDP-17. Nature 393: 702-705.
8. Davis RL, Shrimpton AE, Holohan PD et al. (1999) Familial dementia caused by polymerization of mutant neuroserpin. Nature 401: 376-379.
9. Harada A, Oguchi K, Okabe S et al. (1994) Altered microtubule organization in small-calibre axons of mice lacking tau protein. Nature 369: 488-491.
10. Ikegami S, Harada A, Hirokawa N (2000) Muscle weakness, hyperactivity, and impairment in fear conditioning in tau-deficient mice. Neurosci Lett 279: 129-132.
11. Götz J, Probst A, Spillantini MG et al. (1995) Somatodendritic localization and hyperphosphorylation of tau protein in transgenic mice expressing the longest human brain tau isoform. EMBO J 14: 1304-1313.
12. Brion JP, Tremp G, Octave JN (1999) Transgenic expression of the shortest human tau affects ist compartmentalization and its phosphorylation as in the pretangle stage of Alzheimer's disease. Am J Pathol 154: 255-270.
13. Sturchler-Pierrat C, Abramowski D, Duke M et al. (1997) Two amyloid precursor protein transgenic mouse models with Alzheimer disease-like pathology. Proc Natl Acad Sci U S A 94: 13287-13292.
14. Probst A, Götz J, Wiederhold KH et al. (2000) Axonopathy and amyotrophy in mice transgenic for human four-repeat tau protein, Acta Neuropathol 99: 469-481.
15. Götz J, Nitsch RM (2000) Compartmentalized and cell type-dependent phosphorylation of tau in transgenic mice, submitted
16. Gallyas F (1971) Silver staining of Alzheimer's neurofibrillary changes by means of physical development. Acta Morphol Acad Sci Hung 19: 1-8.
17. Rouleau GA, Clark AW, Rooke K et al. (1996) SOD1 mutation is associated with accumulation of neurofilaments in amyotrophic lateral sclerosis. Ann Neurol 39: 128-131.
18. Ishihara T, Hong, M, Zhang, B et al. (1999) Age-dependent emergence and progression of a tauopathy in transgenic mice overexpressing the shortest human tau isoform. Neuron 24: 751-762.

19. Spittaels K, van den Haute C, van Dorpe J et al. (1999) Prominent axonopathy in the brain and spinal cord of transgenic mice overexpressing four-repeat human tau protein. Am J Pathol 155: 2153-2165.
20. Götz, J, Barmettler R, Ferrari A et al. (2000 In vivo analysis of wild-type and FTDP-17 tau mutant mice. NY Acad Sci, in press.
21. Shockett P, Difilippantonio M, Hellman N et al. (1995) A modified tetracycline-regulated system provides autoregulatory, inducible gene expression in cultured cells and transgenic mice. Proc Natl Acad Sci U S A 92: 6522-6526.
22. Hasegawa M, Crowther RA, Jakes R et al. (1997) Alzheimer-like changes in microtubule-associated protein Tau induced by sulfated glycosaminoglycans. Inhibition of microtubule binding, stimulation of phosphorylation, and filament assembly depend on the degree of sulfation. J Biol Chem 272: 33118-33124.
23. Braak H, Braak E (1995) Staging of Alzheimer's disease-related neurofibrillary changes. Neurobiol Aging 16: 271-284.
24. Sontag E, Nunbhakdi-Craig V, Bloom GS et al. (1995) A novel pool of protein phosphatase 2A is associated with microtubules and is regulated during the cell cycle. J Cell Biol 128: 1131-2244.
25. Götz J, Probst A, Ehler E et al. (1998) Delayed embryonic lethality in mice lacking protein phosphatase 2A catalytic subunit Calpha. Proc Natl Acad Sci U S A 95: 12370-12375.
26. Götz J, Kues W (1999) The role of protein phosphatase 2A catalytic subunit Calpha in embryogenesis: evidence from sequence analysis and localization studies. Biol Chem 380: 1117-1120.
27. Götz J, Probst A, Mistl C et al. (2000) Distinct role of protein phosphatase 2A subunit Cα in the regulation of E-cadherin and β-catenin during development, Mech Dev 93: 83-93.
28. Kins S, Nitsch R, Götz J (2000) Distinct tau hyperphosphorylation in mice with neuronal expression of a dominant negative mutant form of protein phosphatase 2A, in preparation

REGULATION OF ALZHEIMER β-AMYLOID PRECURSOR TRAFFICKING AND METABOLISM

[1,2]Samuel Gandy and [1]Suzana Petanceska
[1]Department of Psychiatry, New York University, The Nathan S. Kline Institute for Psychiatric Research Orangeburg NY 10962, and [2]Fisher Center for Alzheimer Research The Rockefeller University New York NY 10021

1. INTRODUCTION

Alzheimer's disease (AD) is characterized by the intracranial accumulation of the 4 kDa amyloid-ß peptide (Aß), following proteolysis of a ~700-amino acid, integral membrane precursor, the Alzheimer amyloid precursor protein (APP). The best evidence causally linking APP to Alzheimer's disease has been provided by the discovery of mutations within the APP coding sequence that segregate with disease phenotypes in autosomal dominant forms of familial Alzheimer's disease (FAD). Though FAD is rare (<10% of all AD), the hallmark features, amyloid plaques, neurofibrillary tangles, synaptic and neuronal loss, neurotransmitter deficits, dementia, are indistinguishable when FAD is compared with typical, common, "non-familial", or sporadic, AD (SAD). Studies of some clinically relevant mutant APP molecules from FAD families have yielded evidence that APP mutations can lead to enhanced generation or aggregability of Aß, consistent with a pathogenic role in AD. Other genetic loci for FAD have been discovered which are distinct from the immediate regulatory and coding regions of the APP gene, indicating that defects in molecules other than APP can also specify cerebral amyloidogenesis and FAD. To date, all APP and non-APP FAD mutations can be demonstrated to have the common feature of promoting amyloidogenesis of Aß.

Epidemiological studies indicate that postmenopausal women on

Neuropathology and Genetics of Dementia, Edited by Tolnay and Probst
Kluwer Academic/Plenum Publishers, New York 2001

estrogen replacement therapy (ERT) have their relative risk of developing SAD diminished by about one-third as compared with age-matched women not receiving ERT[1]. Because of the key role of cerebral Aß accumulation in initiating AD pathology, it is most attractive that estradiol might modulate SAD risk or age-at-onset by inhibiting Aß accumulation. A possible mechanistic basis for such a scenario is reviewed here.

2. ALZHEIMER'S DISEASE IS ASSOCIATED WITH INTRACRANIAL AMYLOIDOSIS

Amyloid is a generic description applied to a heterogeneous class of tissue protein precipitates, which have the common feature of beta-pleated sheet secondary structure, a characteristic which confers affinity for the histochemical dye Congo red. Amyloids may be deposited in a general manner throughout the body (systemic amyloids) or confined to a particular organ (e.g., cerebral amyloid, renal amyloid). AD is characterized by clinical evidence of cognitive failure in association with cerebral amyloidosis, cerebral intraneuronal neurofibrillary pathology, neuronal and synaptic loss, and neurotransmitter deficits. The cerebral amyloid of AD is deposited around meningeal and cerebral vessels, as well as in gray matter. In gray matter, the deposits are multifocal, coalescing into miliary structures known as plaques. Parenchymal amyloid plaques are distributed in brain in a characteristic fashion, differentially affecting the various cerebral and cerebellar lobes and cortical laminae.

The main constituent of cerebrovascular amyloid was purified and sequenced by Glenner and Wong in 1984. This 40-42 amino acid polypeptide, designated "ß protein" (or, according to Masters and colleagues, "A4"; now standardized as "Aß" by The Husby Commission), is derived from a 695-770 amino acid precursor, termed the amyloid precursor protein (APP), which was characterized by molecular cloning using oligonucleotide probes from the published amyloid peptide sequence. The primary citations for this molecular neuropathology review can be found elsewhere[2].

3. Aß IS A CATABOLITE OF AN INTEGRAL PRECURSOR

The deduced amino acid sequence of APP predicts a protein with a single transmembrane domain. Isoform diversity is generated by alternative mRNA splicing. Isoforms of 751 and 770 amino acids include a protease inhibitor domain ("Kunitz-type protease inhibitor" domain, or KPI) in the extracellular region of the APP molecule. The ectodomains of the protease inhibitor-bearing isoforms of APP are identical to molecules previously identified according to their tight association with proteases, and thus were designated "protease nexin II" ("PN-II"). Identical molecules are also present in the platelet α-granules, where they were described under the name of "factor XI_a inhibitor" (XI_aI). Upon degranulation of the platelet, factor XI_aI (PN-II/APP) exerts an antiproteolytic effect on activated factor XI_a at late steps of the coagulation cascade. Some evidence suggests that KPI-lacking isoforms may also act as regulators of proteolysis. While many bioactivities for APP have been catalogued in the literature, Aß might simply be a by-product of APP metabolism which is not necessarily intimately involved with APP function.

The proteolytic processing steps for APP have by now been definitively identified by purification and sequencing. The first to be identified involves cleavage within the Aß domain. A large amino-terminal fragment of the APP extracellular domain ("protease nexin-II", "PN-II"; or s-APPα or $APPs_\alpha$, for soluble APP_α) is released into the medium of cultured cells and into the cerebrospinal fluid, leaving a small nonamyloidogenic carboxyl-terminal fragment associated with the cell. This pathway is designated the α-secretory cleavage/release processing pathway for APP, because the enzyme(s) which perform(s) this nonamyloidogenic cleavage/release has been designated α-secretase. Thus, one important processing event in the biology of APP acts to preclude amyloidogenesis by proteolyzing APP within the Aß domain, between residues Aß16 and Aß17.

Until mid-1992, the prevailing notion was that Aß production might be restricted to the brain and perhaps even then only in association with aging and AD. This concept became obsolete with the discovery by several groups that a soluble Aß species (presumably a forerunner of the aggregated fibrillar species, which is deposited in senile plaque cores) is detectable in body fluids from various species and in the conditioned medium of cultured cells. Still, Aß, "soluble Aß" (to distinguish it from deposited Aß amyloid) is not readily detectable in the lysates of most cultured cells. Soluble Aß is apparently generated in several cellular compartments, including the endoplasmic reticulum, the Golgi apparatus, the *trans*-Golgi network, and endosomes. Aß production might begin by cleavage at the Aß amino-terminus by another enzyme, designated ß-secretase, probably beginning in the constitutive secretory pathway, i.e., the *trans*-Golgi network (TGN). This model accounts especially well for providing a codistribution opportunity involving APP and ß-secretase prior to APP's encounter with the Aß-domain-destroying α-secretase, which appears to act primarily at the plasma membrane. Data from one FAD mutant (the "Swedish" FAD mutant) support this model strongly. Other data indicate that at least some Aß arises from APP molecules following their residence at the cell surface; presumably these APP molecules have escaped cell-surface α-secretase and have been internalized intact, encountering ß-secretase in an endosome. An Aß variant bearing Glu^{11} at its N-terminus has been recently noted as a prominent product of APP metabolism in cultured neurons[3]. The generation and pharmacologic regulation of [Glu^{11}] Aß formation appears to parallel that observed for standard [Asp^{1}] Aß1-x. The cellular and biochemical origin of the alternative cleavage at Glu^{11} and its role in amyloidogenesis, if any, are all yet to be elucidated.

It is important to appreciate that one area of particular current interest is the elucidation of the molecular and cellular mechanisms by which the carboxyl-terminus of Aß is generated, since this region of the APP molecule resides within an intramembranous domain. The protease(s) responsible for that cleavage, designated the γ-secretase(s) are particularly interesting not only because of the novelty of the active site-substrate reaction (i.e., within a membranous domain) but also owing to evidence that another class of FAD (known as presenilin-related FAD[4]) appears to act via control of γ-secretase function[5,6]. Along this line, the Aß C-terminus is heterogeneous, being composed mostly of peptides terminating at residue Aß40; however, a small but important minority of peptides terminate at residue Aß42, and it is these highly aggregatable peptides that are believed to initiate Aß accumulation in all forms of the disease[7,8].

The primary citations for the foregoing review of APP processing can also be found elsewhere[2].

4. PATHOGENIC APP MUTATIONS AND ALZHEIMER-LIKE PHENOTYPES IN TRANSGENIC MOUSE MODELS *IN VIVO*

Certain mutations associated with familial cerebral amyloidoses have been identified within or near the Aß region of the coding sequence of the APP gene. These mutations segregate with the clinical phenotypes of either hereditary cerebral hemorrhage with amyloidosis, Dutch type (HCHWAD, or FAD-Dutch) or more typical familial Alzheimer's disease ("British", "Indiana" FAD), and provide support for the notion that aberrant APP metabolism is a key feature of AD.

In FAD-Dutch, an uncharged glutamine residue is substituted for a charged glutamate residue at position 22 of Aß. This mutated residue is located in the extracellular region of APP, within the Aß domain, where it apparently exerts its proamyloidogenic effect by generating Aß molecules which bear enhanced aggregation properties and perhaps also by altering α-secretase cleavage. A "Flemish" form of FAD involves a mutation at position 21 of Aß and generates a phenotype intermediate between the relatively pure vascular amyloidosis of "Dutch FAD" and the more typical FAD involving parenchymal and vascular amyloidosis.

Mutations in APP, which are apparently pathogenic for more typical FAD have also been discovered. In the first discovered FAD mutation, an isoleucine residue is subsituted for a valine residue within the transmembrane domain, at a position two residues downstream from the carboxyl-terminus of the Aß domain, equivalent to "Aß44" (if such a position existed). Although a conservative substitution, the mutation segregates with FAD in pedigrees of American, European ("British"), and Asian origins, arguing against the possibility that the mutations represent irrelevant polymorphisms. Other pedigrees have been discovered in which affected members have either phenylalanyl ("Indiana") or glycyl residues at position "Aß44". Neuropathological examination has verified the similarity of these individuals to typical SAD neuropathology. These "Aß44" mutant APPs are the most common of the FAD-causing APP mutations, and the mechanism by which the mutations exert their effects appears to be by enhancing generation of hyperaggregable, C-terminally extended Aß peptides (especially Aß42).

Another FAD pedigree has been established involving, a large Swedish kindred. In this instance, tandem missense mutations occur just upstream of the amino terminus of the Aß domain. Transfection of cultured cells with APP molecules containing the "Swedish" missense mutations results in the production of six- to eight-fold excess soluble Aß above that generated from wild-type APP. This is the first (and, to date, only) example of Alzheimer's disease apparently caused by excessive Aß production. An important issue for clarification in sporadic Alzheimer's disease will be to establish whether hyperaggregation or hyperproduction of Aß (or neither) is/are important predisposing factor(s) to this much more commonly encountered clinical entity. The primary citations for the foregoing review of the genetic-neuropathological correlations of APP mutant FAD can be found elsewhere[2].

Some of the most exciting and newest evidence in support of the pathogenicity of these mutants has arisen from the recent demonstration that certain mutant APP transgenic mice exhibit not only cerebral amyloidosis but also neurofibrillary tangle epitopes and neuronal loss[9]. If these animals also meet key behavioral criteria, Koch's postulates will have been met for the re-creation of all key features of Alzheimer's disease in a species not otherwise susceptible to the illness.

5. SIGNAL TRANSDUCTION REGULATES APP PROCESSING

As work continues toward molecular discovery of α-, ß-, and γ-secretase enzymes, substantial progress has been made toward understanding the regulation of APP cleavage. For example, the relative utilization of the various alternative APP processing pathways appears to be at least partially cell-type determined, with transfected AtT20 cells secreting virtually all APP molecules whereas glia release little or none. In neuronal-like cells, the state of differentiation plays a role in determining the relative utilization of the pathways, with the differentiated neuronal phenotype being associated with relatively diminished basal utilization of the nonamyloidogenic α-secretase cleavage/release pathway.

It is also well accepted that certain signal transduction systems involving protein phosphorylation are important regulators of APP cleavage, acting in many cases by stimulating the relative activity of nonamyloidogenic cleavage by α-secretase. The role of protein kinase C (PKC) and PKC-linked first messengers has received the most attention. In many types of cultured cells, activation of PKC by phorbol esters dramatically stimulates APP proteolysis[10] and $sAPP_{\alpha}$ cleavage/release[11-13]. PKC-stimulated α-secretory cleavage of APP may also be induced by the application of neurotransmitters and other first messenger compounds whose receptors are linked to PKC[14,15]. Okadaic acid, an inhibitor of protein phosphatases 1 and 2A, also increases APP proteolysis and release via the α-secretase pathway[4,15]. Indeed, it has recently been shown in primary neuronal cultures that protein phosphatase inhibition by okadaic acid is much more effective in lowering Aß production than is PKC activation[3]. Thus, either stimulation of PKC or inhibition of protein phosphatases 1 and 2A is sufficient to produce a dramatic acceleration of nonamyloidogenic APP degradation (not shown). PKC-activated APP processing can be demonstrated to diminish generation of Aß in virtually every situation tested to date, including primary neuronal cultures and rodent brain *in vivo*[16-18]. This phenomenon is sometimes referred to as "regulated cleavage of APP" or "reciprocal regulation of APP metabolism"; i.e., when certain signals are activated, α-secretase cleavage increases and ß- and Glu^{11}-secretase cleavages decrease[3].

Steroid hormones such as 17ß-estradiol[19-21] and dihydroepiandrostenedione (DHEA)[22] are other signal transduction compounds which can apparently regulate APP metabolism *in vitro*. Several of these investigators[20,21] have demonstrated that estradiol diminishes Aß generation, while others[19,22] have documented accumulation of sAPP in the conditioned media of cultured cells treated with estradiol[19] or DHEA[22]. One attractive explanation for the accumulation of sAPP is that gonadal steroids might stimulate either the amount or the activity of α-secretase. This explanation is consistent with the finding that the cell-associated product of this activity (the 9-kDa nonamyloidogenic carboxyl-terminal fragment) is increased by estrogen treatment[19]. For estradiol in particular, one means by which the activity of α-secretase might increase is by an increased in the level or activity of PKC, leading to increased "regulated APP cleavage" along the pathway outlined above. Estrogen has been shown to regulate protein kinase C in both normal and neoplastic tissue and to potentiate the phorbol ester-stimulated release of luteinizing hormone from cultured pituitary cells. The effects of estrogen on protein kinase C might be mediated by polypeptide growth factors, many of which activate protein kinase C and a number of which have been demonstrated to enhance the accumulation of sAPP in the conditioned media of cultured cells. The colocalization of estrogen receptors in the basal forebrain, cerebral cortex, and hippocampus with the receptors for nerve growth factor, as well as with nerve growth factor itself, is consistent with the idea that growth factors might play a role in mediating the effects of estrogen in the central nervous system. A review of the literature associating estradiol with protein kinase maybe found elsewhere[19].

Extending this line of reasoning, a more complex relationship involving intercellular signals and APP metabolism is currently evolving: both steroid hormones and neurotrophins have been demonstrated to potentiate the ability of incoming neurotransmitters to activate the "regulated cleavage of APP", as measured by sAPP release[23-27]. The "regulated cleavage" model makes it likely that when compounds from either or both of these classes of substances (i.e., gonadal steroids, neurotrophins) are present at the time of neurotransmitter application, the fold-diminution of Aß and [Glu^{11}] Aß will be greatly enhanced. Assessment of this model is currently underway.

6. INSIGHTS INTO MECHANISM(S) OF REGULATED APP PROCESSING

In general terms, the possible mechanisms for activated processing of integral molecules can be conceptualized as involving either activation or redistribution of either the substrate (i.e., APP) or the enzyme (i.e., α-secretase). Based on APP cytoplasmic tail mutational analyses[28,29], the "substrate activation" model[30] is inadequate to explain activated processing of APP, since abrogation of APP cytoplasmic tail phosphorylation fails to abolish PKC-regulated α-cleavage. In order to test the model of PKC-regulated APP redistribution, Xu et al. demonstrated that the PKC-stimulated biogenesis of nascent transport vesicles from the *trans*-Golgi network includes vesicles, which bear [$^{35}SO_4$] APP[31]. These vesicles would be predicted to fuse with the plasma membrane where codistribution of APP and α-secretase would take place, leading to generation of $sAPP_\alpha$ at the plasma membrane and release of that $sAPP_\alpha$ into the culture medium. Given the preponderant localization of APP within the TGN[32], it is likely that PKC-regulated TGN vesicle biogenesis plays a role in regulating APP metabolism, although its importance cannot be completely assessed until the identification of all relevant PKC targets of this apparatus has been completed.

Along a related line of investigation, Bosenberg et al.[33], succeeded in demonstrating faithful activated processing of the TGF-α precursor in porated cells in the virtual absence of cytosol, and in the presence of N-ethylmaleimide or 2.5 M NaCl. The preservation of activated processing under such conditions suggests that extensive vesicular trafficking is unlikely to be required for all forms of activated processing, and supports a model of direct or indirect enzyme activation which involves an intrinsic plasma membrane mechanism (e.g., direct phosphorylation of an α-secretase or its regulatory subunit). Cell biological and molecular cloning data indicate that APP α-secretases include, or are related to, the metalloproteinase family of adamlysins, including TGFα and pro-tumor necrosis factor convertases. These enzymes appear to have overlapping specificities for these substrates. Several candidate secretases have recently been successfully cloned, and these data should facilitate the resolution of the issue of which enzymes are most important for cleaving which substrates[34-37]. Aspartyl proteinases are the most important α-secretases in yeast[37], and their potential roles, if any, in mammalian APP metabolism, remain to be elucidated.

7. THERAPEUTIC MANIPULATION OF Aß GENERATION VIA LIGAND OR HORMONAL MANIPULATIONS

From a therapeutic standpoint, then, the modulation of APP metabolism via signal transduction pharmacology might be beneficial in individuals with, or at risk for, AD (reviewed in[2,13,18,38]), and it is along this line that 17ß-estradiol, via its possible ability to elevate PKC activity (reviewed in[19]), came to be identified as a potential modulator of Aß metabolism. Estradiol might also modify other factors contributing to Aß deposition and fibril formation, including the processing of soluble Aß into an aggregated form and/or the association of Aß with other molecules, such as α-1-antichymotrypsin, heparan sulfate proteoglycan, and apolipoprotein E[39]. In particular, it has recently been recognized that apolipoprotein E is an important participant in Aß accumulation[40], and that apolipoprotein E isoforms play important roles as differential determinants of risk for AD[41]. Human postmortem[42] and cell culture[43] data are also beginning to indicate that apolipoprotein E isoforms exert at least some of their effects via controlling Aß accumulation, perhaps at the stage of clearance of Aß peptides. Apolipoprotein E-isoform-specific synaptic remodeling also appears to be a consistent and potentially relevant observation[44].

Events beyond Aß deposition may also be crucial in determining the eventual toxicity of Aß plaques. Aggregation of Aß is important for *in vitro* models of neurotoxicity, but the relevance of this phenomenon for the pathogenesis of AD is unclear, since Aß deposits may occur in normal aging, in the absence of any evident proximate neuronal injury[45]. This suggests that other events must distinguish "diffuse" cerebral amyloidosis with wispy deposits of Aß alone from "full-blown" AD with neuritic plaques, and neuron and synapse loss. One intriguing possible contributing factor is the association of complement components with Aß. In cerebellum, where Aß deposits appear to cause no injury, plaques are apparently free of associated complement, while in the forebrain, complement associates with plaques, perhaps becoming activated and injuring the surrounding cells[46]. Other, as-yet undiscovered, plaque-associated molecules may also play important roles.

8. ESTROGEN REGULATES BRAIN Aß LEVELS IN EXPERIMENTAL ANIMALS

The majority of AD (over 90%) is sporadic, and the identification of factors that influence the onset and/or progression of the disease would be an important step toward understanding its mechanism(s) and for developing successful, rational therapies. Along this line, compelling evidence indicates that estrogen status plays an important role in the etiology of the disease, since postmenopausal women receiving estrogen replacement therapy (ERT) have a significantly delayed or reduced risk of developing AD[5,6]. A recent review of 15 case-control and cohort studies reported since 1990 indicates that postmenopausal women on ERT halve their risk of developing AD[3].

In light of these observations, and since physiological concentrations of estrogen were shown to decrease the levels of Aß released from rodent or human primary neuronal cultures[7], we recently investigated the ability of female gonadal hormone status to modulate brain Aß levels *in vivo*.

To investigate the effects of ovariectomized (ovx) and E2 replacement on brain Aß levels, 4 groups of guinea pigs were studied: an intact group (n=8), an ovariectomized group (ovx; n=9), an ovariectomized group treated with 1mg of E2 per kg of body weight per day, (ovx+low E2; 1mg/kg BW/day; n=9), and an ovariectomized group treated with 5mg of E2 per kg of body weight per day (ovx+high E2; 5mg/kg BW/day; n=8).

Ovx resulted in decreased serum E2 levels as compared to age-matched, intact animals (Figure 1A). Ten days of replacement with low dose E2 (1mg E2/kg BW/day) or high dose E2 (5mg E2/kg BW/day) led to dose-dependent increases in serum E2 levels when compared to either the ovx group or to the intact group (Figure 1A). The values for serum E2 levels of the intact animals varied: some were comparable to the serum E2 levels of ovx animals, while others were comparable to the serum E2 levels of ovx+low-dose E2 animals (Figure 1A). This variation is typical of the normal asynchronous cycling of intact animals[47]. The high-dose E2 treatment resulted in supraphysiological levels of serum E2 (Figure 1A).

The uteri of the ovx animals were hypotrophic when compared to the uteri of the intact group of animals: on average, uteri from ovx animals weighed less than one third that of uteri from intact animals (Figure 1B). The uteri of the ovx animals that had received low-dose E2 for 10 days were hypertrophied and had weights comparable to, or higher than, those of the intact group (Figure 1B). High-dose E2 treatment was also associated with uterine hypertrophy, though the uterine weights did not exceed those of the ovx+low-dose E2 group (Figure 1B).

The 10-week ovx was also associated with increased levels of brain Aß as compared to intact animals (1.5-fold average increase in total Aß; $p<0.0001$) (Figure 2A). It is of note that the levels of Aß42 increased to a greater extent than the levels of Aß40 (1.8-fold average increase for Aß42; $p<0.0001$, and 1.5-fold average increase for Aß40; $p<0.00001$) (Figure 2B, 2C). This resulted in an increased Aß42/Aß40 ratio in the ovx group as compared to the intact group (1.3-fold average increase; $p<0.001$).

Treatment of ovx guinea pigs with the low-dose E2 for 10 days, beginning 8 weeks after ovx, was associated with significant reversal of the ovx-induced elevation of total brain Aß levels (18% average decrease; $p<0.01$) (Figure 2A). Aß40 and Aß42 levels decreased to a similar extent (18% average decrease for Aß40, $p<0.01$; 21% average decrease for Aß42, $p=0.033$) (Figure 2B, C). The high-dose E2 treatment (5mg/kg BW/day) had a similar effect on brain Aß levels and did not cause any additional decrease in either Aß species (Figure 2). Interestingly, in a few individual animals receiving E2, the levels of brain Aß were similar to, or lower than, those observed in animals from the intact group (Figure 2B, 2). Since both E2 treatments lowered Aß40 and Aß42 levels similarly, the Aß42/Aß40 ratios were similar between the two groups, and on average, not significantly lower than the average Aß42/Aß40 ratio for the ovx group. However, it is of note that the Aß42/Aß40 ratios for a few individual animals from the E2 treatment groups fell within the range of ratios observed in animals from the intact group (not shown). No apparent correlation between serum estradiol levels and brain Aß levels was observed in any of the experimental groups.

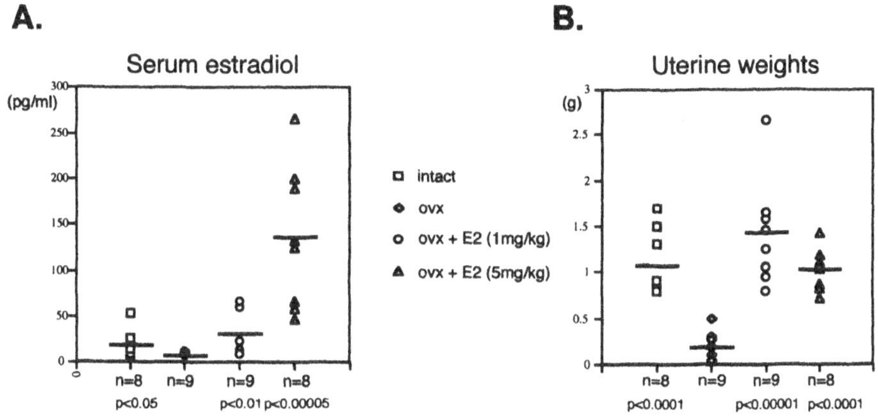

Figure 1. Effect of ovariectomy and E2 treatment on serum estradiol levels (1A) and uterine weight (1B).

The levels of $sAPP_\alpha$ were unaffected by ovx or E2 replacement (Figure 3). This effect on $sAPP_\alpha$ is in contrast with data from cell culture studies where the estrogen-induced decrease in Aß peptides was accompanied by an increase in $sAPP_\alpha$ levels in the cell culture media[20]. Our findings are consistent with those of another report showing that *in vivo* treatment of rodent brain with phorbol ester caused decreased Aß release but had no effect on $sAPP_\alpha$ levels[18]. Taken together, these two independent sets of observations suggest that the reciprocal relationship between Aß peptide and $sAPP_\alpha$ release that has been observed in cultured cells[20] may be less evident or absent in brain *in vivo*.

These are the first data indicating that the levels of Aß in the brains of living mammals are under the control of gonadal hormones. More specifically, we present evidence that prolonged ovariectomy is associated with increased brain Aß40 and Aß42 levels, and that this increase can be significantly reversed by E2 treatment. Thus, ovx guinea pigs may be a useful system for evaluating the impact of estrogen derivatives, estrogen-like compounds and selective estrogen receptor modulators on brain Aß metabolism *in vivo*.

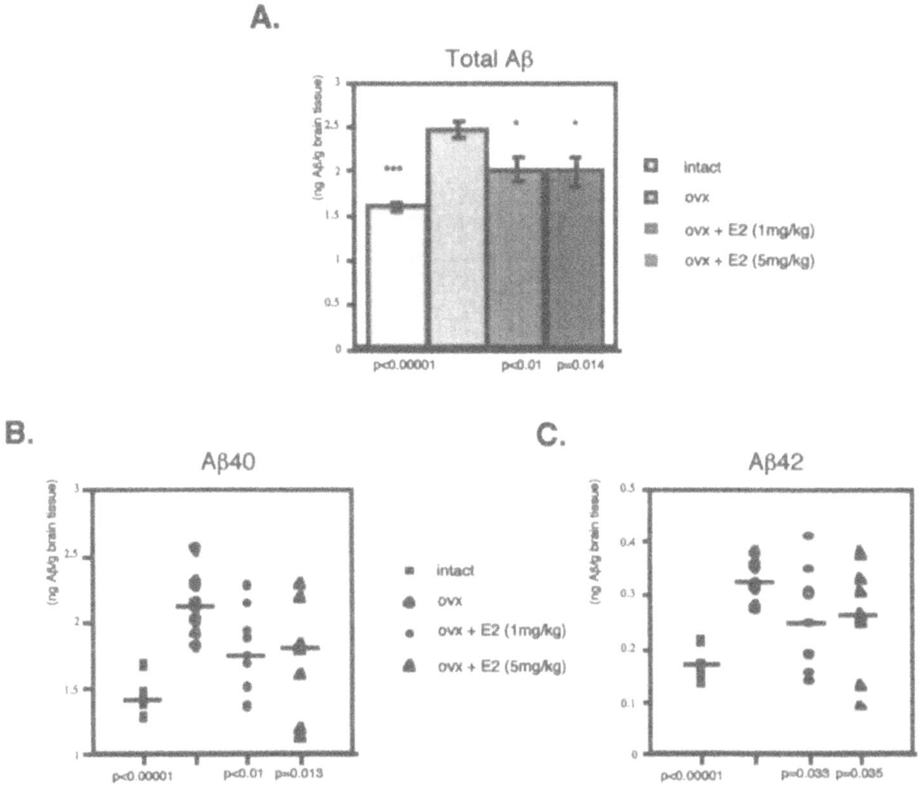

Figure 2. Effect of ovariectomy and E2 replacement on Aß levels.

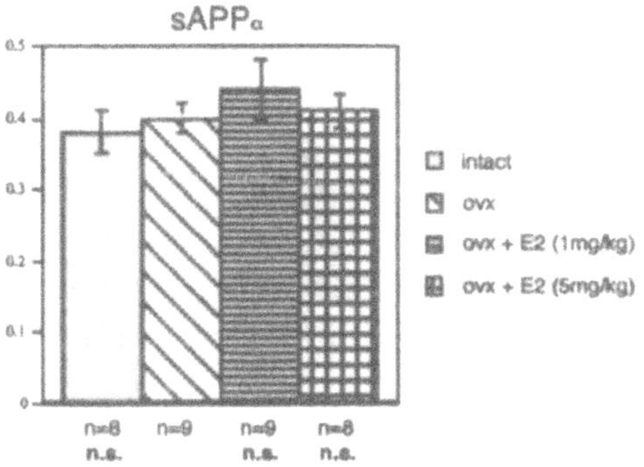

Figure 3. Effect of ovariectomy and E2 treatment on sAPPα levels in brain.

Since our studies involved assays of steady state levels of APP metabolites in response to ovariectomy and E2 replacement, we are unable to distinguish whether the changes in Aß levels reflect altered Aß generation or altered Aß clearance. Also, it remains to be determined whether the observed effects on Aß metabolism occur in response to activation of brain estrogen receptors or whether they are mediated by classical, estrogen receptor-independent mechanisms.

Cessation of ovarian estrogen production in postmenopausal women might facilitate Aß deposition by increasing the local concentrations of Aß40 and Aß42. A related study on plaque-forming transgenic mice, showing that prolonged ovx accelerates the elevation of brain Aß levels at a time when amyloid deposition is incipient, supports this hypothesis (Karen Duff and Lorenzo Refolo, personal communication). Our finding that estrogen treatment is associated with diminution of brain Aß levels suggests that modulation of Aß metabolism may be one of the ways by which estrogen prevents and/or delays the onset of AD in postmenopausal women.

It remains possible that the estrogen-associated preservation of cognitive function in post-menopausal women results from multiple activities of estrogen. In addition to modulating Aß accumulation, these other activities of estrogen include: providing trophic support for basal forebrain cholinergic neurons[48], stimulating neurite outgrowth and synaptogenesis[49], stimulating apolipoprotein E expression[50,51] and/or protecting neurons from oxidative stress and Aß-induced toxicity[52]. The availability of complementary *in vivo* systems will enable the investigation of each of these neuroactivities of estrogen under physiological (i.e., guinea pigs) and pathophysiological (i.e., plaque-forming transgenic mice) conditions, and will facilitate the experimental dissection of the relative contributions of each of these activities.

REFERENCES

1. Tang MX, Jacobs D, Stern Y et al. (1996) Effect of estrogen during menopause on risk and age at onset of Alzheimer's disease. Lancet 348: 429-432.
2. Gandy S, Greengard P (1994) Processing of Aß-amyloid precursorprotein: Cell biology, regulation, and role in Alzheimer disease. Int Rev Neurobiol 36: 29-50.
3. Gouras GK, Xu H, Jovanovic JN et al. (1998) Generation and regulation of beta-amyloid peptide variants by neurons. J Neurochem 71: 1920-1925.
4. Sherrington R, Rogaev EI, Liang Y et al. (1995) Cloning of a gene bearing missense mutations in early-onset familial Alzheimer's disease. Nature 375: 754-760.
5. Scheuner D, Eckman C, Jensen M et al. (1996) Secreted amyloid beta-protein similar to that in the senile plaques of Alzheimer's disease is increased in vivo by the presenilin 1 and 2 and APP mutations linked to familial Alzheimer's disease. Nat Med 2: 864-870.
6. Borchelt DR, Thinakaran G, Eckman C et al. (1996) Familial Alzheimer's disease-linked presenilin 1 variants elevate Abeta 1-42/1-40 ratio in vitro and in vivo. Neuron 17: 1005-1013.

7. Iwatsubo T, Odaka A, Suzuki N et al. (1994) Visualization of Abeta 42(43) and Abeta 40 in senile plaques with end-specific Abeta monoclonals: evidence that an initially deposited species is Abeta 42(43). Neuron 12: 45-53.
8. Lemere CA, Blusztajn JK, Yamaguchi M et al. (1996) Sequence of deposition of heterogeneous amyloid beta-peptides and *APOE* in Down syndrome: implications for initial events in amyloid plaque formation. Neurobiol Dis 3: 16-32.
9. Calhoun ME, Wiederhold KH, Abramowski D et al. (1998) Neuron loss in APP transgenic mice. Nature 395: 755-756.
10. Buxbaum JD, Gandy SE, Licchetti P et al. (1990) Processing of Alzheimer ß/A4 amyloid precursor protein: Modulation by agents that regulate protein phosphorylation. Proc Natl Acad Sci USA 87: 6003-6006.
11. Caporaso GL, Gandy SE, Buxbaum JD et al. (1992) Protein phosphorylation regulates secretion of Alzheimer ß/A4 amyloid precursor protein. Proc Natl Acad Sci USA 89: 3055-3059.
12. Gillespie SL, Golde TE, Younkin SG (1992) Secretory processing of the Alzheimer amyloid ß/A4 protein precursor is increased by protein phosphorylation.. Biochem Biophys Res Commun 187: 1285-1290.
13. Sinha S, Lieberburg I (1992) Normal metabolism of the amyloid precursor protein (APP). Neurodegeneration 1: 169-175.
14. Buxbaum JD, Oishi M, Chen HI et al. (1992) Cholinergic agonists and interleukin I regulate processing and secretion of the Alzheimer ß/A4 amyloid protein precursor. Proc Natl Acad Sci USA 89: 10075-10078.
15. Nitsch RM, Slack BE, Wurtman RJ et al. (1992) Release of Alzheimer amyloid precursor derivatives stimulated by activation of muscarinic acetylcholine receptors. Science 258: 304-307.
16. Buxbaum JD, Koo EH, Greengard P (1993) Protein phosphorylation inhibits production of Alzheimer amyloid ß/A4 peptide. Proc Natl Acad Sci USA 90: 9195-9198.
17. Hung AY, Haass C, Nitsch RM et al. (1993) Activation of protein kinase C inhibits cellular production of the amyloid ß-protein.. J Biol Chem 268: 22959-22962.
18. Savage MJ, Trusko SP, Howland DS et al. (1998) Turnover of amyloid beta-protein in mouse brain and acute reduction of its level by phorbol ester. J Neurosci 18: 1743-1752.
19. Jaffe AB, Toran-Allerand CD, Greengard P et al. (1994) Estrogen regulates metabolism of Alzheimer amyloid beta precursor protein.. J Biol Chem 269: 11306-11308.
20. Xu H, Gouras GK, Greenfield JP et al. (1998) Estrogen reduces neuronal generation of Alzheimer beta-amyloid peptides. Nat Med 4: 447-451.
21. Chang D, Kwan J, Timiras PS (1997) Estrogens influence growth, maturation and amyloid beta-peptide production in neuroblastoma cells and in a beta-APP transfected kidney 293 cell line. Adv Exp Med Biol 429: 261-271.
22. Danenboerg HD, Haring R, Fisher A et al. (1996) Dehydroepiandrosterone (DHEA) increases production and release of Alzheimer's amyloid precursor protein. Life Sci 59: 1651-1657.
23. Danenberg HD, Haring R, Heldman E et al. (1995) Dehydroepiandrosterone augments M1-muscarinic receptor-stimulated amyloid precursor protein secretion in desensitized PC12M1 cells. Ann NY Acad Sci: 774: 300-303.

24. Fisher A, Heldman E, Garwitz D et al. (1996) M1 agonists for the treatment of Alzheimer's disease: Novel therapeutic properties and clinical update. Ann NY Acad Sci 777: 189-196.
25. Haring R, Fisher A, Marciano D et al. (1998) Mitogen-activated protein kinase-dependent and protein kinase C-dependent pathways link the m1 muscarinic receptor to ß-amyloid precursor protein secretion. J Neurochem 71: 2094-2103.
26. Roßner S, Ueberham U, Schliebs R et al. (1998) p75 and trkA receptor signalling independently regulate amyloid precursor protein mRNA expression, isoform composition and protein secretion in PC12 cells. J Neurochem 71: 757-766.
27. Roßner S, Ueberham U, Schliebs R et al. (1998) The regulation of amyloid precursor protein metabolism by cholinergic mechanisms and neurotrophic receptor signalling. Prog Neurobiol 56: 541-569.
28. da Cruz e Silva 0, da Cruz e Silva AO, Zaia CT et al. (1993) Regulated cleavage of Alzheimer ß-amyloid precursor protein in the absence of the cytoplasmic tail. Neuroscience 57: 873-877.
29. Sahasrabudhe SR, Spryt MA, Muenkel HA et al. (1992) Release of amino-terminal fragments from amyloid precursor protein reporter and mutated derivatives in cultured cells. J Biol Chem 267: 25062-25608.
30. Gandy S, Czernik AJ, Greengard P (1988) Phosphorylation of Alzheimer disease amyloid precursor peptide by protein kinase C and Ca^{+2}/calmodulin-dependent protein kinase II. Proc Natl Acad Sci USA 85: 6218-6221.
31. Xu, H, Greengard P, Gandy S (1995) Regulated formation of Golgi secretory vesicles containing Alzheimer beta-amyloid precursor protein. J Biol Chem 270: 23243-23245.
32. Caporaso G, Takei K, Gandy SE et al. (1994) Morphologic and biochemical analysis of the intracellular trafficking of the Alzheimer ß/A4 amyloid precursor protein. J Neurosci 14: 3122-3138.
33. Bosenberg MW, Pandiella A, Massague J (1993) Activated release of membrane-anchored TGF-α in the absence of cytosol. J Cell Biol 122: 95-101.
34. Black RA, Rauch CT, Kozlosky CJ et al. (1997) A metalloproteinase disintegrin that releases tumour-necrosis factor-alpha from cells. Nature 385: 729-733.
35. Peschon JJ, Slack JL, Redaly P et al. (1998) An essential role for ectodomain shedding in mammalian development. Science 282: 1281-1284.
36. Buxbaum JD, Lin KN, Luo Y et al. (1998) Evidence that tumor necrosis factor alpha converting enzyme is involved in regulated alpha-secretase cleavage of the Alzheimer amyloid protein precursor. J Biol Chem 273: 27765-27767.
37. Komano H, Seeger M, Gandy SE et al. (1998) Involvement of cell surface glycosyl-phosphatidylinositol-linked aspartyl proteases in alpha-secretase-type cleavage and ectodomain solubilization of human Alzheimer beta-amyloid precursor protein in yeast. J Biol Chem 273: 31648-31651.
38. Gandy S, Greengard P (1992) Amyloidogenesis in Alzheimer's disease: Some possible therapeutic opportunities. Trends Pharmacol Sci 13: 108-113.
39. Wisniewski T, Frangione B (1992) Apolipoprotein E: a pathological chaperone in patients with cerebral and systemic amyloid. Neurosci Lett 135: 235-238.
40. Bales KR, Verina T, Dodel RC et al. (1997) Lack of apolipoprotein E dramatically reduces amyloid beta-peptide deposition. Nat Genet 17: 263-264.
41. Strittmatter WJ, Saunders AM, Schmechel D et al. (1993) Apolipoprotein E: High-avidity binding to ß-amyloid and increased frequency of type 4 allele in late-onset familial Alzheimer disease. Proc Natl Acad Sci USA 90: 1977-1981.

42. Rebeck GW, Reiter JS, Strickland DK et al. (1993) Apolipoprotein E in sporadic Alzheimer's disease: allelic variation and receptor interactions. Neuron 11: 575-580.
43. Yang DS, Small DH, Seydel U et al. (1999) Apolipoprotein E promotes the association of amyloid-ß with Chinese Hamster Ovary cells in an isoform-specific manner. *Neuroscience* 90: 1217-1226
44. Sun Y, Wu S, Bu G et al. (1998) Glial fibrillary acidic protein-apolipoprotein E (apoE) transgenic mice: astrocyte-specific expression and differing biological effects of astrocyte-secreted apoE3 and apoE4 lipoproteins. J Neurosci 18: 3261-3172.
45. Masliah E, Terry RD, Alford M et al. (1990) Diffuse plaques do not accentuate synapse loss in Alzheimer's disease. Am J Pathol 137: 1293-1297.
46. Lue LF, Rogers J (1992) Full complement activation fails in diffuse plaques of the Alzheimer's disease cerebellum. Dementia 3: 308-313.
47. Shi F, Ozawa M, Komura H et al. (1999) Secretion of ovarian inhibin and its physiologic roles in the regulation of follicle-stimulating hormone secretion during the estrous cycle of the female guinea pig. Biol Reprod 60: 78-84.
48. Luine VN (1985) Estradiol increases choline acetyltransferase activity in specific basal forebrain nuclei and projection areas of female rats. Exp Neurol 89: 484-490.
49. McEwen BS, Woolley C (1994) Estradiol and progesterone regulate neuronal structure and synaptic connectivity in adult as well as developing brain. Exp Gerontol 29:431-436.
50. Srivastava RA, Srivastava N, Averna M et al. (1997) Estrogen up-regulates apolipoprotein E (ApoE) gene expression by increasing ApoE mRNA in the translating pool via the estrogen receptor alpha-mediated pathway. J Biol Chem 27: 33360-33366.
51. Stone DJ, Rozovsky I, Morgan TE et al. (1997) Astrocytes and microglia respond to estrogen with increased apoE mRNA in vivo and *in vitro*. Exp Neurol 143:313-318.
52. Gridley KE, Green PS, Simpkins JW (1998) A novel, synergistic interaction between 17ß-estradiol and glutathione in the protection of neurons against Aß 25-35-induced toxicity In vitro. Mol Pharmacol 4: 874-880.

ROLE OF TYPE 10 17β-HYDROXYSTEROID DEHYDROGENASE IN THE PATHOGENESIS OF ALZHEIMER'S DISEASE

Song-Yu Yang and Xue-Ying He
Department of Pharmacology, New York State Institute for Basic Research in Developmental Disabilities, Staten Island, New York 10314

Human type 10 17β-hydroxysteroid dehydrogenase (17β-HSD10) is a multifunctional enzyme[1]. It displays L-3-hydroxyacyl-CoA dehydrogenase activity[2] and belongs to the short chain dehydrogenase family[3], and thus was previously designated as short chain L-3-hydroxyacyl-CoA dehydrogenase (SCHAD). This protein has a high affinity for amyloid-β peptide and was reportedly localized at the endoplasmic reticulum, and for that reason it was formerly known as endoplasmic reticulum-associated amyloid-β peptide-binding protein (ERAB)[4]. We provide evidence that 17β-HSD10 is a nuclear gene-encoded mitochondrial enzyme. It is the only brain 17β-HSD known as yet able to modulate intracellular estrogen levels in a brain region susceptible to AD pathology. The findings suggest that disturbance of steroid hormone homeostasis by high levels of brain 17β-HSD10, rather than its binding of Aβ, underlies the involvement of this protein in Alzheimer's disease.

1. HUMAN *SCHAD* GENE ENCODING 17β-HSD10

Kobayashi et al. isolated a new mitochondrial L-3-hydroxyacyl-CoA dehydrogenase from bovine liver[5]. We cloned the cDNA for its human counterpart and isolated recombinant human brain SCHAD from *E. coli*[2]. Human brain SCHAD is a homotetramer with a molecular mass of 108 kDa[2].

Neuropathology and Genetics of Dementia, Edited by Tolnay and Probst
Kluwer Academic/Plenum Publishers, New York 2001

We found that this novel L-3-HAD is a new member of the short chain dehydrogenase family[2]. As shown in Figure 1, the three-dimensional structure of human 17β-HSD10 is very similar to those of other short chain dehydrogenases, such as 3α/20β-HSD from *Streptomyces exfoliatus* and 7α-HSD from *E. coli*. The designation SCHAD reflected this important feature as well as distinguished it from members of the L-3-hydroxyacyl-CoA dehydrogenase (L-3-HAD) family[6,7]. The primary structure of SCHAD is identical to that of ERAB[2].

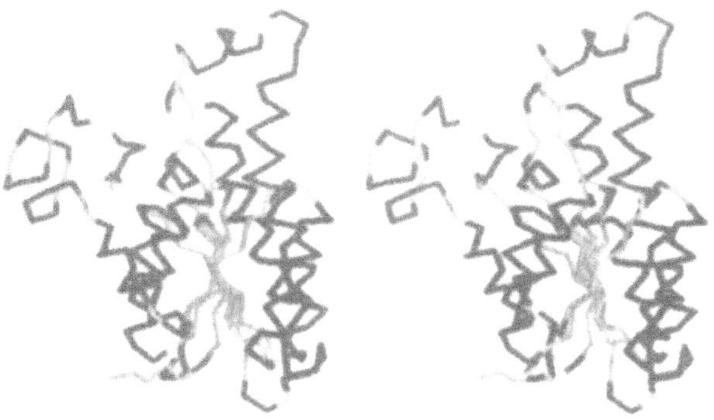

Figure 1. Stereodiagram of the α-carbon model of type 10 17β-hydroxysteroid dehydrogenase. The subunit is a single-domain dehydrogenase with an open twisted α/β structure.

Therefore, ERAB is encoded by human *SCHAD* gene, which is organized into six exons and five introns and maps to chromosome Xp11.2 (Figure 2)[2].

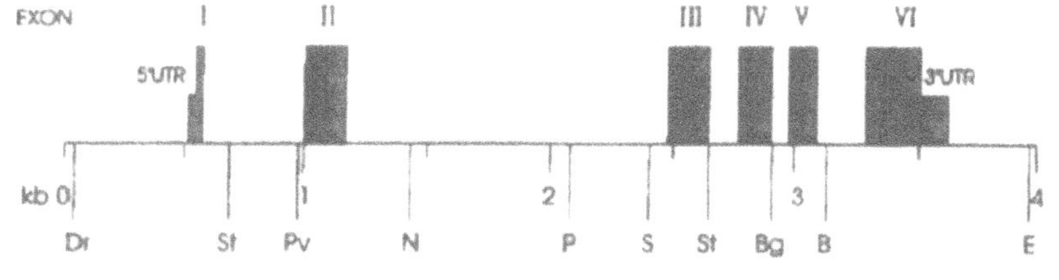

Figure 2. Structural organization and restriction map of human *SCHAD* gene encoding type 10 17β-hydroxysteroid dehydrogenase. At the top of the scaled diagram are Roman numerals above the boxes indicating the positions of exons. The shorter boxed regions are 5'- and 3'-untranslated regions (UTR) as indicated. The numbers below the line indicate the length of the gene in kilobases. Positions of restriction enzyme recognition sites are designated: Dr, DrdII; St, StuI; Pv, PvuII; N, NsiI; P, PstI; S, SphI; Bg, BglI; B, BamHI; E, EcoRI.

The amino-terminal NAD-binding region of the dehydrogenase is encoded by the first three exons, whereas the remaining exons code for the carboxyl-terminal substrate-binding region harboring putative catalytic residues[2]. We found that the *SCHAD* gene is the first gene of a cluster of six genes that reportedly escape chromosome X inactivation[8]. We found that the gene ADS13 is about 5 kb downstream relative to the *SCHAD* gene, which is identical to the ADS9 gene, while the IB772 gene maps at about 98 kb upstream of the *SCHAD* gene. The order of this multigene domain on chromosome X was thus clarified (Figure 3). Therefore, the control of the *SCHAD* gene expression appears to be a key to the puzzle why this multigene domain could escape chromosome X inactivation. If the *SCHAD* gene on either chromosome X belonging to a female is active, there may be a gender-based difference in expression levels of 17β-HSD10.

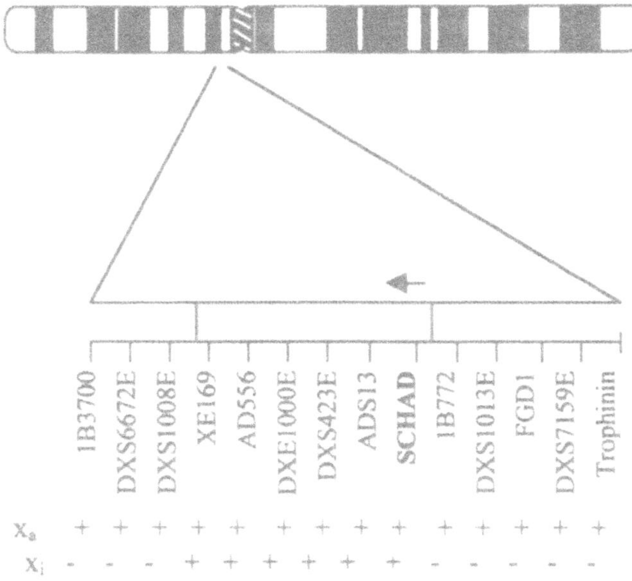

Figure 3. Human *SCHAD* gene expression from active (X_a) and inactive (X_i) chromosome X. The *SCHAD* gene is the first of six genes that escape inactivation and map within a domain which is transcriptionally active (+) on the inactive X chromosome. Genes mapping both distal and proximal to this domain (six genes) are transcriptionally silent (-) on the inactive X chromosome.[8] An arrow indicates the transcription direction of the *SCHAD* gene.

2. CATALYTIC FUNCTIONS OF 17β-HSD10

As a member of the short chain dehydrogenase family[3]. SCHAD/ERAB was expected to have a rather broad substrate specificity. It can oxidize L-3-hydroxyacyl-CoAs with various acyl chain lengths[2], and also catalyzes the redox reactions of various steroid hormones[1,9-11]. Catalytic rate constants reported for ERAB proved to be overestimated[11]. SCHAD displays appreciable 17β/3α-HSD activities, and its catalytic efficiency for the oxidation of 17β-estradiol is comparable to that of 17β-HSD4[1,12], which also harbors D-3-hydroxyacyl-CoA dehydrogenase activity[12]. SCHAD/ERAB was thus redesignated as type 10 17β-HSD, a new member of the 17β-HSD/17KSR family[13-15]. Its catalytic functions related to sex steroid metabolism are shown in Table 1. In contrast to other types of 17β-HSDs,[13,14] 17β-HSD10 oxidizes 5α-androstane-3α,17β-diol (3α-Adiol) with a relatively high catalytic efficiency to 5α-dihydrotestosterone (5α-DHT), but not to androsterone. Also, the oxidative 3α-HSD activity of 17β-HSD10 effectively converts androsterone to androstanedione. Since both 5α-DHT ($K_i = 6x10^{-8}$ M) and androstanedione ($1x10^{-9}$ M< K_i < $6x10^{-8}$ M) are very strong inhibitors of brain aromatase[16], they can suppress the biosynthesis of estrogen from testosterone. Obviously, high levels of 17β-HSD10 will reduce intracellular levels of 17β-estradiol. 17β-Estradiol reduces neuronal Aβ generation[17], and increases the expression of Bcl-XL to prevent Aβ-induced apoptosis[18]. The protective and trophic effects of 17β-estradiol on neurons are well documented[19]. Therefore, overexpression of 17β-HSD10 will increase the vulnerability of neurons to Aβ or oxidant stresses.

Substrate	**Product**	K_m (μM)	V_{max} (nmol•min^{-1}•mg^{-1})	K_{cat} (min^{-1})	K_{cat}/K_m (min^{-1}•mM^{-1})
17β-Estradiol	Estrone	43 ± 2.1	24.9 ± 0.53	0.66 ± 0.01	15.3
Estrone	17β-Estradiol	—	—	ND	—
5α-Dihydro-testosterone	3α-Andro-stanediol	112 ± 18	71.9 ± 7.69	1.94 ± 0.21	17.3
3α-Andro-stanediol	5α-Dihydro-testosterone	34 ± 2.4	207 ± 6.30	5.58 ± 0.17	164
Andro-stanedione	Androsterone	44 ± 2.5	8.7 ± 0.40	0.23 ± 0.01	5.2
Androsterone	Andro-stanedione	45 ± 9.3	24.4 ± 2.97	0.66 ± 0.08	14.7

Table 1. Kinetic Parameters of human type 10 17β-Hydroxysteroid Dehydrogenase (adapted from references [1,9]. ND, not detectable

3. EXPRESSION LEVELS OF 17β-HSD10

It was reported that SCHAD mRNA is expressed in gonadal and peripheral tissues at different levels[9]. Levels of 17β-HSD10 in different human tissues were determined by immunoblotting (Figure 4). This enzyme can function in two different lipid metabolic pathways[1]. It is well known that muscle uses fatty acids as an important energy source[6]. However, it was reported that only bovine mitochondrial L-3-HAD, but not SCHAD (HAD II), was detected in skeletal muscle by immunoblotting[5]. Although we recently found a small amount of SCHAD in skeletal muscle, it is certainly much lower than amounts found in other tissues (Figure 4). This observation suggests that L-3-HAD, but not SCHAD, is essential to the fatty acid β-oxidation pathway. In contrast to muscle, the fatty acid β-oxidation flow rate in brain is extremely low[20]. Since L-3-HADs are present in great excess than required to support this rate of fatty acid oxidation[20], the function of 17β-HSD10 (SCHAD) in brain is likely to be maintenance of intracellular sex steroid levels.

Figure 4. Western blot analysis of human SCHAD levels in different tissues. Proteins (50μg/lane) were separated by SDS-PAGE on a 4-20% gradient gel at pH 8.3 and then the separated proteins were transferred to a nitrocellulose membrane by semi-dry blotting. After blocking the blot, proteins were probed with rabbit anti-17βHSD10 (1: 500) followed by reaction with secondary antibodies (goat anti-rabbit IgG conjugated with alkaline phosphatase, BioRed). Antigen bands were visualized by the development of the blot in a staining mixture containing the alkaline phosphatase substrate. Lanes 1, heart; 2, brain; 3, placenta; 4, lung; 5, liver; 6, skeletal muscle; 7, kidney; 8, ovary; 9, testis.

17β-HSD type 2 is not expressed in brain temporal lobe[21], and human 17β-HSD type 4 in brain is expressed only in Purkinje cells[22]. In contrast, 17β-HSD10 (ERAB/SCHAD) is abundantly expressed in brain temporal lobe as well as other regions of human brain[4]. It was also reported that AD brains have significantly more 17β-HSD10 (ERAB/SCHAD) than age-matched controls[4]. Type 10 is the only 17β-HSD known to play a part in estrogen homeostasis in the brain area susceptible to AD pathology. High 17β-HSD10 activities would cause an intracellular 17β-estradiol-deficient state and attenuate estrogen's neuroprotective benefits.

4. INTRACELLULAR LOCALIZATION OF 17β-HSD10

ERAB was reported to be associated with the ER[4]. However, we showed that SCHAD is located in mitochondria[1,9]. A recent ERAB report has claimed that ERAB has a dual localization, both ER and mitochondria[10]. ERAB was considered as a putative intracellular Aβ receptor[4,10,23,24]. If this protein is localized in mitochondria only, it seems unlikely that this protein would serve as a receptor to mediate the neurotoxicity of Aβ[4], or to form an ERAB-Aβ complex interfering with the axonal transport of proteins[23], or to play a substantial part in ER-specific apoptosis as proposed previously[24]. It has been announced that a signal sequence is absent in ERAB suggesting that it is bound to cytoplasmic aspect of the ER[4,23]. On the other hand, Sambamurti and Lahiri reported that ERAB contains a putative noncleavable signal peptide that leads ERAB to bind to Aβ in the lumen of ER and transport vesicles[25]. Evidently, it is of great importance to find the truth about the location of 17β-HSD10 (SCHAD/ERAB). In order to unequivocally solve the controversy about the intracellular localization of this important protein, there is a fundamental question that must be answered. What is the structural basis of this protein for its transport to mitochondria? We identified a noncleavable mitochondrial signal sequence at the N-terminal of 17β-HSD10 (SCHAD/ERAB) by generating a NTS-GFP fusion protein harboring the N-terminal eleven amino acid residues of human[2] or rat[15] 17β-HSD10 (Met-Ala-Ala-Ala-Cys-Arg-Ser-Val-Lys-Gly-Leu) at its N-terminus. Wild type GFP spreads over the cytoplasm and nucleus whereas the NTS-GFP fusion protein yields green granular images. The intracellular distribution pattern of NTS-GFP is quite distinct from that of GFP but well co-localized with mitochondria. The results demonstrate that the N-terminal sequence of human or rat type 10 17β-HSD can lead a reporter protein (GFP) to be specifically transported to mitochondria, to the exclusion of other locations. The mitochondrial targeting sequence of 17β-HSD10 is highly homologous to the presequence of the precursor of mitochondrial L-3-hydroxyacyl-CoA dehydrogenase with the potential for a positively charged amphiphilic helix (cf Figure 5)[9,26-27]. Point mutations R6E/K9N or a deletion mutation (Δ2-7) was introduced to destroy the N-terminal mitochondrial targeting signal. However, the truncated mutant 17β-HSD10 was still imported to mitochondria, but not other organelles such as the ER. The intracellular distribution pattern of the 17β-HSD10 (R6E/K9N) mutant without any net positive charges in its N-terminal region was also observed to be identical to that of wild type 17β-HSD10 (see Figure 5 of reference[9]). The results suggest that 17β-HSD10 has a downstream mitochondrial importing signal in addition to the N-terminal mitochondrial targeting signal. Since none of the mutants in which the prominent mitochondrial targeting signal has been destroyed were found to be associated with the ER, the presence of a possibly weak ER targeting signal that could explain the reportedly dual localization has been indisputably excluded.

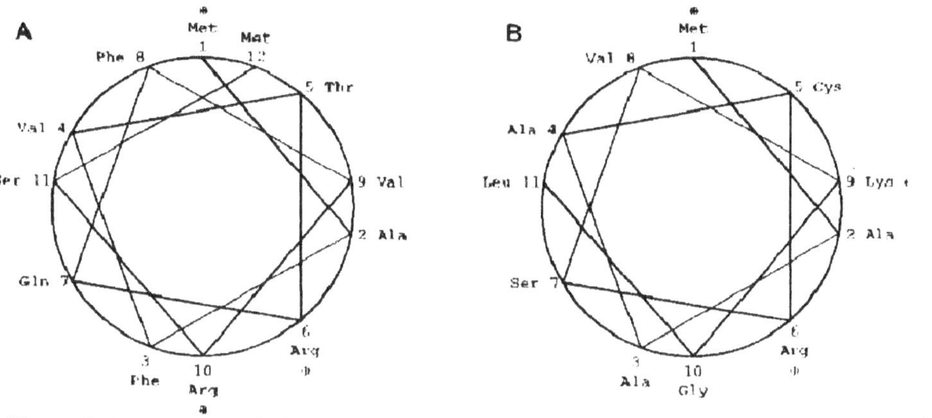

Figure 5. Presequence of pig mitochondrial L-3-hydroxyacyl-CoA dehydrogenase7 (A) and the mitochondrial targeting sequence of 17β-HSD10 from human[2] or rat[15] (B) displayed as 'helical wheels' (3.6 residues per turn). They are similar amphiphilic helices.

The ERAB-Aβ binding theory was supported by a subcellular fractionation experiment[4,10]. However, it must first exclude the possibility that ER fractions in ERAB studies have been contaminated with mitochondria through a modification of a conventional fractionation protocol[28]. Cell homogenate was separated to the clarified supernatant and pellet by centrifugation. Organelle contents and 17β-HSD10 levels in cell homogenate, clarified supernatant, and the resuspended pellet were examined by immunoblotting analysis. The clarified supernatant was loaded on a series of sucrose steps to make enriched ER fractions as described in the conventional fractionation procedure[28] while a resuspension of the pellet was loaded on another sucrose steps to be fractionated as described in the modified fractionation procedure employed for ERAB studies[10]. It was found that 17β-HSD10 levels in the supernatant and the ER fraction prepared by the conventional protocol were too low to be detected. In contrast, both 17β-HSD10 and a mitochondrial marker enzyme, cytochrome oxidase (COX), were found to be abundant in the resuspended pellet as well as in the ER fraction prepared by the modified protocol[10]. The key difference between the conventional[28] and the modified[10] fractionation protocols is that "the latter material", namely the resuspended pellet, rather than the clarified supernatant was used as the starting material for the modified subcellular fractionation procedure employed in ERAB studies[10]. Our subcellular fractionation data corroborate the reported results from immunocytochemistry-confocal microscopy studies[9]. A body of evidence leads to the conclusion that 17β-HSD10 is a nuclear gene-encoded mitochondrial enzyme. The localization of 17β-HSD10 suggests that brain mitochondria not only play an essential role in steroidogenesis, 20,22-desmolase (CYP11A1) is located exclusively in mitochondria so that the initial reaction in the biosynthesis of steroid hormones must take place in mitochondria[29], but mitochondria also likely play an important part in modulating intracellular levels of the active forms of sex steroid hormones.

5. CONCLUDING REMARKS

Since this protein is, in fact, not associated with the ER as previously reported, it is very unlikely that 17β-HSD10 (ERAB) serves as a putative Aβ receptor at the endoplasmic reticulum to mediate Aβ neurotoxicity or plays a role in ER-specific apoptosis. High levels of 17β-HSD10 in mitochondria would reduce the intracellular levels of estrogen. The reported increase of Aβ-induced apoptosis in cells with an overexpression of 17β-HSD10 (ERAB) is probably due to an increase of the neurons' vulnerability to Aβ stress. The hypothesis that disturbance of sex steroid hormone homeostasis in a brain region susceptible to AD pathology by high levels of 17β-HSD10 is strongly supported by all the experimental data currently available. The replacement of the theory that ERAB is a putative Aβ receptor at the ER by a theory based on the catalytic functions of 17β-HSD10 paves the way for exploring the role(s) of this multifunctional protein in the pathogenesis of Alzheimer's disease.

ACKNOWLEDGEMENTS

We are grateful to Dr. D. Lin for constructing the α-carbon model of type 10 17β-HSD. This work was supported in part by U.S. Public Health Service Grant AG04220 from the National Institute on Aging and by the Alzheimer's Association and the New York State Office of Mental Retardation and Developmental Disabilities.

REFERENCES

1. He XY, Merz G, Mehta P et al. (1999) Human brain short chain L-3-hydroxyacyl coenzyme A dehydrogenase is a single-domain multifunctional enzyme. Characterization of a novel 17β-hydroxysteroid dehydrogenase. J Biol Chem 274: 15014-15019.
2. He XY, Schulz H, Yang SY (1998) A human brain L-3-hydroxyacyl coenzyme A dehydrogenase is identical to an amyloid β-peptide binding protein involved in Alzheimer's disease. J Biol Chem 273: 10741-10746.
3. Jornvall H, Persson B, Krook M et al. (1995) Short-chain dehydrogenase/reductases (SDR). Biochemistry 34: 6003-6013.
4. Yan SD, Fu J, Soto C et al. (1997) An intracellular protein that binds amyloid-β peptide and mediates neurotoxicity in Alzheimer's disease. Nature 389: 689-695.
5. Kobayashi A, Jiang LL, Hashimoto T (1996) Two mitochondrial 3-hydroxyacyl-CoA dehydrogenases in bovine liver. J. Biochem. (Tokyo) 119, 775-782 (1996).
6. Yang SY, He XY (1999) Molecular mechanisms of fatty acid β-oxidation enzyme catalysis. In: Current Views of Fatty Acid Oxidation and Ketogenesis: From Orgenelles to Point Mutations (Quant PA, Eaton S, eds), pp 133-143: Kluwer Academic/Plenum Publishers, New York.
7. He XY, Zhang G, Blecha F et al. (1999) Identity of heart and liver L-3-hydroxyacyl-CoA dehydrogenase. Biochim Biophys Acta 147: 119-123.
8. Miller AP, Willard HF (1998) Chromosomal basis of X chromosome inactivation: Identification of a multigene domain in Xp11.21-p11.22 that escapes X inactivation. Proc Natl Acad Sci USA 95: 8709-8714.
9. He XY, Merz G, Yang YZ et al. (2000 Function of human brain short chain L-3-hydroxyacyl coenzyme A dehydrogenase in androgen metabolism. Biochim Biophys Acta 1484: 267-277.
10. Yan SD, Shi Y, Zhu A et al. (1999) Role of ERAB/L-3-hydroxyacyl-coenzyme A dehydrogenase type II activity in Aβ-induced cytotoxicity. J Biol Chem 274: 2145-2156.
11. He XY, Yang YZ, Schulz H et al. (2000) Intrinsic alcohol dehydrogenase and hydroxysteroid dehydrogenase activities of human mitochondrial short chain L-3-hydroxyacyl-CoA dehydrogenase. Biochem J 345: 139-143.
12. Adamski J, Normand T, Leenders F et al. (1995) Molecular cloning of a widely expressed human 80 kDa 17β-hydroxysteroid dehydrogenase IV. Biochem J 311: 437-443.
13. Peltoketo H, Luu-The V, Simard J et al. (1999) 17β-Hydroxysteroid dehydrogenase (HSD)/17-ketosteroidreductase (KSR) family; nomenclature and main characteristics of the 17HSD/KSR enzymes. J Mol Endocrinol 23: 1-11.
14. Su J, Lin M, Na Natl Acad Sci Complementary deoxyribonucleic acid cloning and enzymatic characterization of a novel 17β,3α-hydroxysteroid/retinoid short chain dehydrogenase/reductase. Endocrino USA 95: 8709-87.
15. He XY, Merz G, Yang YZ et al. (2000Molecular cloning, modeling, and localization of rat type 10 17β-hydroxysteroid dehydrogenase. Mol Cell Endocrinol: in press.
16. Wozniak A, Hutchison JB (1993) Action of endogenous steroid inhibitors of brain aromatase relative to fadrozole. J Steroid Biochem Mol Biol 44: 641-645.
17. Xu H, Gouras GK, Greenfield JP et al. (1998) Estrogen reduces neuronal generation of Alzheimer β-amyloid peptides. Nature Medicine 4: 447-451.

18. Pike C (1999) Estrogen modulates neuronal Bcl-xL expression and β-amyloid-induced apoptosis: relevance to Alzheimer's disease. J Neurochem 72: 1552-1563.
19. Dubal DB, Wilson ME, Wise PM (1999) Estradiol: a protective and trophic factor in the brain. Alzheimer's Disease Review 4: 1-9.
20. Yang SY, He XY, Schulz H (1987) Fatty acid oxidation in rat brain is limited by the low activity of 3-ketoacyl-coenzyme A thiolase. J Biol Chem 262: 13027-13032.
21. Stoffel-Wagner B, Watzka M, Steckelbroeck S et al (1999) Expression of 17β-hydroxysteroid dehydrogenase types 1, 2, 3 and 4 in the human temporal lobe. J Endocrinol 160: 119-126.
22. Moller G, Leenders F, van Grunsven EG et al. (1999) Characterization of the HSD17β4 gene: D-specific multifunctional protein 2/17β-hydroxysteroid dehydrogenase IV. J Steroid Biochem Mol Biol 69 441-446 (1999).
23. Beyreuther K, Masters CL (1997) The ins and outs of amyloid-β. Nature 389: 677-678.
24. Nakagawa T, Zhu H, Morishima N et al. (2000) Caspase-12 mediates endoplasmic-reticulum-specific apoptosis and cytotoxicity by amyloid-β. Nature 403: 98-103.
25. Sambamurti K, Lahiri DK (1998) ERAB contains a putative noncleavable signal peptide. Biochem Biophys Res Commun 249: 546-549.
26. Schatz G (1996) The protein import system of mitochondria. J Biol Chem 271: 31763-31766.
27. Neupert N (1997) Protein import into mitochondria. Annu Rev Biochem 66: 863-917.
28. Vidugiriene J, Vidugiriene J (1993) Early lipid intermediates in glycosylphosphatidylinositol anchor assembly are synthesized in the ER membrane bilayer. J Cell Biol 121: 987-996.
29. Simpson ER, Waterman MR. Steroid hormone biosynthesis in the adrenal cortex and its regulation by adrenocorticotropin. In: Endocrinology (DeGroot LJ, ed.), pp. 1630-1641, 3rd Edition, W. B. Saunders, Co., Philadelphia, PA.

THE SPECTRUM OF VASCULAR DISEASE IN DEMENTIA

From ischaemia to amyloid angiopathy.

Roy O. Weller and Stephen D. Preston

Divisions of Cell and Molecular Medicine and Neuroscience, University of Southampton School of Medicine, Southampton SO16 6YD, UK.

1. INTRODUCTION

Age is not only a major risk factor for Alzheimer's disease and other dementias but also for cerebrovascular disease. It has long been recognised that cerebrovascular disease alone may cause brain damage and that it may co-exist with the pathological features of Alzheimer's disease and other dementias[1]. One of the major objectives of the present study is to explore the different ways in which cerebrovascular disease may be involved in the pathogenesis of a variety of dementias, especially Alzheimer's disease.

Cerebral arteries have a dual function; not only do they deliver blood to the brain but, via their perivascular spaces, they also appear to drain interstitial fluid and protein from the brain[2,3,4,5]. Here, we examine how the ageing cerebrovascular system is associated with dementia through (a) thrombo-embolic vascular disease and cerebral infarction, (b) cerebral amyloid angiopathy and the impedence of drainage of beta A4 amyloid from the brain.

Neuropathology and Genetics of Dementia, Edited by Tolnay and Probst
Kluwer Academic/Plenum Publishers, New York 2001

2. THROMBO-EMBOLIC CEREBROVASCULAR DISEASE

Atherosclerosis affects the aorta and major muscular arteries throughout the body. In the larger vessels, such as the internal carotid arteries, ulcerated atherosclerotic plaques are a source of emboli which travel to the brain and result in large or small cerebral infarcts. Intracranial arteries, such as the middle cerebral artery, may be completely occluded by thrombosis associated with rupture of an atherosclerotic plaque; this may result in large areas of cerebral infarction. Coronary artery occlusion resulting in myocardial infarction may also indirectly affect the brain. A mural thrombus in the left ventricle, adherent to an area of myocardial infarction, may be a source of emboli that occlude arteries in the brain, resulting in infarction. Various diseases increase the risk of atherosclerosis; these include diabetes and hypertension. In addition, hypertension is associated with small vessel disease in many organs, including the brain. Intracerebral haemorrhages may result from hypertensive vascular disease but ischaemic lesions are also a complication of hypertension with small infarcts often located in the basal ganglia[6].

As a result of atherosclerotic and hypertensive cerebrovascular disease, brains of elderly individuals may suffer large infarcts producing focal neurological signs. However, the damage to the brain may be more subtle, with multiple small infarcts affecting grey matter in the cerebral cortex and basal ganglia resulting in multi-infarct dementia. Although less common, the white matter may be most severely affected as in the subcortical sclerosis of Binswanger, which again, results in a dementing syndrome[1].

Although much less common than cerebrovascular disease due to atherosclerosis, there are a number of other important causes of ischaemic disease of the brain to consider. Vasculitides such as polyarteritis nodosa, granulomatous arteritis, sarcoidosis and rheumatoid arthritis are uncommon but may all be associated with thrombosis of small vessels in the brain resulting in multiple infarcts and dementia[1]. In some families, CADASIL (cerebral autosomal dominant arteriopathy with subcortical infarcts and leucoencephalopathy) is a major cause of vascular dementia[1,6].

3. CEREBRAL AMYLOID ANGIOPATHY

Alzheimer's disease is characterised pathologically by the intracellular accumulation of neurofibrillary tangles, composed of hyperphosphorylated tau, involving cortical and hippocampal neurons, and by the extracellular accumulation of beta A4 amyloid (Aβ). Histologically stainable Aβ is deposited in diffuse, neuritic and burnt out or cored plaques in the neuropil mainly in the temporal and frontal cerebral cortex and hippocampus. Aβ also accumulates in the walls of blood vessels, both in the cortex and in the leptomeninges, as cerebral amyloid angiopathy (CAA). Occurring in both Alzheimer's disease and in non-demented elderly patients, CAA may be complicated by intracerebral haemorrhage especially in patients over 75 years of age[7]. Although CAA has long been recognised as an important cause of intracerebral haemorrhage, here we concentrate upon the pathogenesis of CAA itself and upon the relationship of CAA to drainage of Aβ from the human brain.

In his seminal publication on cerebral amyloid angiopathy in 1938, Scholz described "drusige Endartung" which refers to globular excrescences attached to blood vessels[8]. Classically, CAA occurs in Alzheimer's disease involving the deposition of Aβ in the walls of intracortical capillaries and arterioles and in leptomeningeal arteries. Vessels in the occipital lobes are particularly frequently involved by CAA but vessels in the frontal and temporal lobes and in other regions of the cerebral cortex may also be affected. Meningeal arteries of the cerebellum may develop CAA, but only rarely are the intraparenchymal cerebellar arteries involved. Within the cerebral cortex, capillaries and arteries of all diameters may show deposition of Aβ, whereas in the leptomeninges, it is mainly the smaller arteries that show deposition of stainable fibrillary Aβ in their walls. Veins are rarely involved[9]. Despite the absence of stainable Aβ in the walls of large intracranial arteries, biochemical studies have shown that Aβ is present in the walls of the middle cerebral artery and the basilar artery even in individuals as young as 30 years of age[10]. Aβ was not, however, recovered from the walls of extracranial arteries[10].

Cerebral amyloid angiopathy is not only a feature of Alzheimer's disease and, to a lesser extent, of non-demented elderly patients, but it also occurs in a number of other amyloidoses. Most prominent of these are the cases of hereditary cerebral haemorrhage with amyloidosis (HCHWA). In the Dutch type (HCHWA-D), an aberrant Aβ is deposited in the walls of cerebral arteries and frequently results in intracerebral haemorrhage[7]. The peptide deposited in in the Icelandic type, HCHWA-I, is cystatin C, but the pattern and complications are similar to the Dutch Aβ type of familial CAA. In Familial British Dementia, amyloid plaques and CAA are prominent features with deposition of ABri in vessel walls in the brain parenchyma. CAA also occurs in which prion protein is deposited in vessel walls as well as in the extracellular space of the brain[11]. The deposition of chemically different amyloids, such as Aβ, cystatin C, and prion protein, suggests that CAA is not unique to Aβ. Thus the different forms of CAA may have a common mechanism in their pathogenesis.

3.1 Pathogenesis of Cerebral Amyloid Angiopathy

A number of hypotheses have been proposed for the pathogenesis of cerebral amyloid angiopathy[9]. Histological, immunocytochemical and electron microscope studies have shown that Aβ is deposited in close relationship to smooth muscle cells in the walls of cortical and leptomeningeal arteries in CAA and that smooth muscle cells in tissue culture produce amyloid precursor protein (APP) from which Aβ is derived[7]. It has, therefore, been proposed that cerebral amyloid angiopathy is the result of the production of Aβ by smooth muscle cells. However, there are a number of factors that militate against this hypothesis. *Firstly,* it is only intracranial blood vessels that are involved in CAA and even the carotid artery in the neck shows no amyloid angiopathy. *Secondly,* CAA involves mainly the smaller arteries containing the least number of smooth muscle cells and the capillaries that lack smooth muscle cells altogether. It seems, therefore, that although production of Aβ by smooth muscle cells may contribute to the overall Aβ load within vessel walls this is not the full explanation for cerebral amyloid angiopathy[9].

It has long been suggested that the amyloid in CAA is derived from the blood[12]. Thus, one of the older names of "dyshoric" angiopathy for vessels in the cortex showing complete replacement of the media by amyloid suggests that the amyloid is derived from the serum. There are, however, problems with this concept. An origin for amyloid peptides from the blood does not explain the similarity in the distribution of different proteins, such as Aβ, cystatin-C and prion proteins, in CAA. Neither can it be easily reconciled with the observation that only intracranial vessels are involved and that small early deposits of amyloid are in the adventitia rather than near the lumen[9]. Only arteries and capillaries, rather than venules, are affected. If proteins were leaking from the serum, extracranial vessels and intracranial veins might be expected to be involved in the dyshoric process.

3.2 Unifying hypothesis for the pathogenesis of cerebral amyloid angiopathy

Any unifying hypothesis for cerebral amyloid angiopathy must take into account that it is not just Aβ that is deposited in vessel walls but a variety of other petides. Thus, CAA may be due to failure of a generic function of blood vessel walls rather than the local production of amyloid proteins within vessel walls themselves.

Here we present evidence for the hypothesis that CAA results from the accumulation of amyloid peptides in perivascular interstitial fluid drainage pathways.

3.3 Drainage of Interstitial Fluid and Proteins from Brain Parenchyma

3.3.1 Experimental studies

Since the classical experiments of Helen Cserr and her co-workers showed that proteins injected into the central grey matter of the rat brain drained to cervical lymph nodes[2], it has been recognised that there is a highly organised system of drainage of interstitial fluid from the brain[5]. Furthermore, the pivotal role played by interstitial fluid drainage from the brain to cervical lymph nodes in B-cell and T-cell mediated immunity of the brain in the rat has been clarified[13,14]. Injection of insoluble tracers into the rat brain has outlined the periarterial pathways by which interstitial fluid drains from the brain[3,15]. By combining the results with a number of experimental studies in the rat and morphological studies in man, a clearer concept of the route taken by fluid and proteins draining from the neuropil has emerged. The extracellular space in grey matter is a narrow, convoluted compartment between neuronal and glial processes[16]. Diffusion of proteins through the extracellular space may depend upon their molecular weight and, in the case of larger proteins and peptides, pathways may be tortuous[16]. The extracellular space, however, is in direct continuity with the perivascular space of capillaries which is mainly occupied by basement membrane, but appears to offer a low resistance pathway for the drainage of fluid and proteins from the extracellular space of the brain[2]. When insoluble particles of Indian ink are injected into the grey matter of the rat brain, they pass along periarterial spaces, both in the cortex and around the major branches of cerebral arteries over the surface of the brain. Mapping the passage of such tracers shows that they pass to the circle of Willis and thence alongside arteries to the olfactory bulbs to enter lymphatics that pass through the cribriform plate into the nasal submucosa and directly to deep cervical lymph nodes[3,5,15].

3.3.2 Studies on the Human Brain

Tracer studies are obviously impractical in man but ultrastructural studies have shown that, in the human brain, the very narrow extracellular space of the neuropil is in direct continuity with the wider perivascular spaces of capillaries. Pericapillary spaces form a direct connection to the pericvascular spaces of cortical arterioles and thence to the much more capacious perivascular spaces of leptomeningeal arteries[17]. However, whereas the extracellular space of the brain is in direct continuity with the capillary perivascular space, a layer of leptomeninges separates the neuropil from perivascular space around muscular arteries in the brain[17]. This layer of leptomeniges, derived from the pia mater, continues to encompass the perivascular space around leptomeningeal arteries and can, indeed, be traced right to the base of the skull around the carotid and vertebral arteries (unpublished observations). In principal, therefore, an anatomical perivascular compartment is present around intracranial blood vessels in man; it extends from capillaries, alongside cortical arterial and leptomeningeal arteries, and even through the base of the skull to the region of cervical lymph nodes in the neck. The physiological significance of this perivascular pathway in man for the lymphatic drainage of interstitial fluid from the human brain is, as yet, unclear. But, as will become apparent from the distribution of beta amyloid in cerebral amyloid angiopathy, Aβ appears to act as a natural tracer for outlining perivascular interstitial fluid drainage pathways in the human brain[9].

3.4 The origin of Aβ in cerebral amyloid angiopathy in Alzheimer's disease.

Beta amyloid (Aβ) is derived from cleavage of the ~700 amino acid long transmembrane protein amyloid precursor protein (APP) via cleavage by the α, β and γ secretases[18]. Aβ is released into the extracellular space following cleavage of APP. Three major forms of Aβ are produced by the action of secretases. The two longer forms of Aβ are peptides of approximately 4kDa; the longer form is 42 or 43 amino acids long [Aβ 1-42(43)] and the shorter form is 40 amino acids long (Aβ 1-40). Due to the presence of hydrophobic amino acids, the long form is less soluble than the shorter soluble form (Aβ 1-40). Furthermore, Aβ may exist as a fibrillar form when deposited in the brain parenchyma and in vessel walls.

There is a pool of soluble Aβ 1-40 in the normal human brain and recent studies have shown that it is the accumulated level of the soluble amyloid in the grey matter in Alzheimer's disease which exceeds the level of soluble amyloid in intellectually normal elderly individuals[19,20]. Senile plaques within the brain consist mainly of Aβ 1-42 but also contain Aβ 1-40 and a number of other components, such as synuclein products, α1-antichymotrypsin, and apolipoprotein E[21], almost as if they were refuse deposits within the extracellular space. Detailed biochemical analyses of intracranial vessels have shown that the more insoluble Aβ 1-42 is deposited first in the walls of cerebral arteries and it is only later that Aβ 1-40 accumulates[22]. This suggests that deposition of Aβ 1-42 may result in the entrapment of Aβ 1-40. It is not just in older individuals that Aβ can be detected in the walls of cerebral blood vessels. Shinkai et al.[10] have shown that, even at the age of 30, Aβ 1-40 and Aβ 1-42 can be recovered from the walls of large intracranial arteries, such as the basilar and middle cerebral arteries. This latter observation strongly supports the concept that amyloid peptides drain along perivascular pathways as a normal physiological function throughout life.

3.5 Distribution of cerebral amyloid angiopathy reflects the drainage pathways for Aβ.

Recent histological and immunocytochemical studies have shown that the pattern of Aβ deposition in CAA correlates well with a perivascular pathway for the drainage of Aβ from the neuropil of the human brain[9], (Figure 1).

As described by Scholz[8] and by many subsequent authors[7], Aβ is deposited in capillary basement membranes from which it may extend focally into the neuropil as smooth globose "Drusen" firmly attached to amyloid in the capillary basement membrane or as more feathery deposits extending in a filiform manner into the neuropil. In the present study, such continuity of Aβ deposits in the neuropil with Aβ in the perivascular space was sėen only around capillaries. Aβ in cortical arterioles may be present in small amounts on the outer aspects of the vessel or as large amounts throughout the vessel wall replacing smooth muscle cells. Although diffuse plaques may be present in the glia limitans around cortical arterioles, such amyloid deposits do not appear to be in continuity with Aβ deposits within the vessel walls themselves. The current study suggests, therefore, that Aβ does not pass directly from the neuropil into the perivascular spaces of arterioles but only into the perivascular spaces of capillaries.

Given the tracer studies in animals, we may conclude that amyloid deposited in the walls of arteries has drained along the perivascular space from the capillaries.

In many cases of Alzheimer's disease, CAA of leptomeningeal vessels is much more prominent than CAA in the underlying cortical vessels[9]. Small deposits of amyloid and, therefore, possibly the earlier deposits in leptomeningeal arteries, tend to occupy the more peripheral parts of the vessel, i.e. related to the perivascular space[9]. More heavily involved vessels show deposition of Aβ throughout the tunica media and between the smooth muscle cells. Electron microscope immunocytochemistry has shown that Aβ is deposited within the basement membrane of the vascular smooth muscle cells[23]. Such deposits may be focal within the vessel wall, giving a banding appearance when the vessels are viewed in longitudinal section[9]. In the more severe cases, smooth muscle cells within the artery wall are completely replaced by Aβ and small aneurysms may form in the regions of the vessels heavily involved by amyloid. Such aneurysms may be the source of cerebral haemorrhage in amyloid angiopathy[7,9].

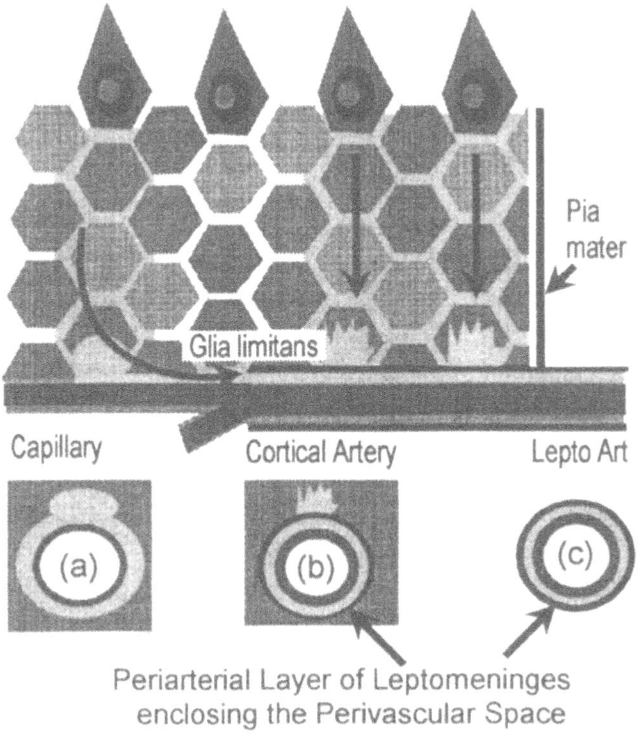

Figure 1. A summary diagram of the proposed drainage pathway for Aβ along vessel walls. *Top Diagram:* Aβ percolates through the narrow extracellular spaces and drains into the pericapillary spaces (curved arrow). Aβ then drains along the periarterial spaces within the cortex to the perivascular channels of the leptomeningeal arteries. Drainage of Aβ from the extracellular compartment of the cortex directly into periarterial spaces appears to be blocked, possibly by the layer of pia mater extending along the vessel. Around arteries, therefore, Aβ accumulates in the glia limitans as well as in the artery wall. *Bottom Diagrams:* Vessels in cross section. Patterns of deposition of Aβ in (a) walls of capillaries - with Druse; (b) wall of an artery and in the perivascular glia limitans; (c) the wall of a leptomeningeal artery.

3.6 Impedence of drainage of Aβ from the brain in Alzheimer's disease - causes and effects.

If, as has been proposed here, CAA in Alzheimer's disease is due to the deposition of Aβ in perivascular interstitial fluid drainage pathways, the question that needs to be answered is why does it accumulate in elderly individuals and not in younger people.

There may be three major factors involved in the impedence of drainage of Aβ along perivascular spaces. *Firstly*, the shape of the drainage pathway. Extending as it does from a high resistance extracellular compartment in the neuropil into a lower resistance capillary perivascular space, the drainage pathway may offer decreasing resistance as the dimensions of the perivascular space increase around arteries in the cortex and in the leptomeninges. This would suggest that interstitial fluid and proteins flow from high resistance to low resistance compartments. To some extent, this has been demontrated in the pattern of distribution of insoluble tracers injected into the rat brain[3,5]. Indian ink injected into the central grey matter of the rat brain passes into the perivascular spaces of the small arteries and preferentially extends into the perivascular space of the larger arteries. Although tracer is seen in the pericapillary spaces, it is small in amount compared with the tracer around larger arteries. Such a reduction in resistance may lead to slowing of flow and deposition of insoluble Aβ, especially in the walls of leptomeningeal arteries which are the vessels most commonly affected by CAA.

The *second* factor is age changes in artery walls. Although there is good evidence for the perivascular drainage of fluid and proteins from the brain in experimental animals and circumstantial evidence for such drainage from the study of CAA in man, the motive force for such drainage is unknown. However, the close association of Aβ deposition with smooth muscle cells and the perivascular space suggests that vessel pulsation may play a role as a motive force for the drainage of interstitial fluid and proteins in the opposite direction to the flow of blood within the artery lumen. Decreasing resistance within the perivascular space from capillary to leptomeningeal artery would also favour such a reverse flow. If arterial pulsations do play a role as the motive force for interstitial fluid drainage along perivascular pathways, loss of elasticity and rigidity of arterial walls may be a factor in impeding the drainage of Aβ in the elderly. One of the most common features of cerebral atherosclerosis is the pipestem rigid ectatic nature of the basilar, vertebral and terminal internal carotid arteries[6]. We compared aged arteries with arteries from younger individuals and showed a loss of elasticity of the vessel wall (unpublished data). Histological sections of the basilar artery from children and young adults shows little fibrosis in the vessel media and a convoluted elastic lamina in the artery in its relaxed state. This reflects the capacity for the young artery to expand during systole and recoil during diastole. Such pulsations are almost certainly lost in the elderly arteriosclerotic artery; the lumen of the aged artery remains dilated and constant and histology shows subintimal fibrosis, stretching of the elastic lamina and thinning of the media. The coincidence of age range of such arteriosclerotic changes and CAA suggests a causal connection may exist, but direct experimental proof has, as yet, not been possible.

Just as the causal mechanisms for cerebral amyloid angiopathy can only be derived from circumstantial evidence, the effects of CAA on brain physiology can only be surmised at present. Loss of muscle cell in vessels severely affected by cerebral amyloid angiopathy may have an effect on autoregulation and reactivity of the cerebral circulation in affected individuals and deposition of amyloid in the perivascular drainage pathways may be a factor in the accumulation of amyloid within the brain. Since the description of amyloid plaques in the brain in Alzheimer's disease and in elderly individuals, numerous studies have concentrated upon the effect of such plaques on neural tissue[1]. The neurotoxicity of amyloid has been investigated and the neurodestructive effects of microglial activation around senile plaques has been studied. Both these factors may play important roles in destruction of neurons in Alzheimer's disease but it has also been recognised recently that the level of soluble Aβ is higher in Alzheimer's disease brains than in non-demented, age matched, elderly patients[19,20]. In effect, the level of soluble amyloid in the brain correlates better with cognitive decline than does the level of insoluble fibrillar amyloid and the number of senile plaques[19]. It is possible, therefore, that a threshold may eventually be reached in elderly patients whereby a combination of cerebrovascular disease of large arteries and the presence of insoluble amyloid deposited in CAA may ultimately impede the drainage of Aβ and contribute to the onset of dementia.

4. CONCLUSIONS

Cerebrovascular disease and Alzheimer's disease in the elderly may have a number of causal relationships. The association of thromboembolic disease and multi-infarct dementia with Alzheimer's disease is well recognised. Smooth muscle cell destruction in leptomeningeal and intracortical arteries in CAA may not only predispose to intracerebral haemorrhage, but may also have an effect on cerebral blood flow and autoregulation. Another factor emphasised here is the effect of CAA on impeding the drainage of interstitial fluid and peptides such as soluble Aβ from the brain. Therapeutic strategies that reduce multi-infarct dementia have been in place for many years, particularly for the treatment of hypercholesterolaemia and hypertension. Strategies for facilitating the drainage of Aβ along perivascular pathways may bear fruit for future therapeutic intervention in patients with Alzheimer's disease. In particular, the identification of chaperone molecules for Aβ and substances that increase the solubility and decrease fibrilogenesis of Aβ may have beneficial effects on this one aspect of Alzheimer's disease. The introduction of APP transgenic mice which develop CAA[24] may greatly facilitate the study of Aβ elimination from the brain by perivascular pathways and the development of therapies that enhance such drainage.

ACKNOWLEDGEMENTS

This work was supported by the Multiple sclerosis Society of Great Britain and Northern Ireland, the David Gibson Fund, the Brain Fund and the Wolfson Foundation (SDP).

REFERENCES

1. Esiri MM, Hyman BT, Beyreuther K et al. (1997) Ageing and dementia. In: Greenfield's Neuropathology (Graham DI, Lantos PL, eds), pp 153-233, 6th ed. Vol II: Arnold, London.
2. Cserr HF, Harling-Berg CJ, Knopf PM (1992) Drainage of brain extracellular fluid into blood and deep cervical lymph and its immunological significance. Brain Pathol 2: 269-276.
3. Zhang ET, Richards HF, Kida S et al. (1992) Directional and compartmentalised drainage of interstitial fluid and cerebrospinal fluid from the rat brain. Acta Neuropathol 83: 233-239.
4. Weller RO (1995) Fluid compartments and fluid balance in the central nervous system. In: Gray's Anatomy (Williams PL, ed), pp 1202-1224, 38th ed.: Churchill Livingstone, Edinburgh .
5. Weller RO (1998) Pathology of cerebrospinal fluid and interstitial fluid of the CNS: significance for Alzheimer disease, prion disorders and multiple sclerosis. J Neuropathol Exp Neurol 57: 885-894.
6. Kalimo H, Kaste M, Haltia M (1997) Vascular disease. In: Greenfield's Neuropathology (Graham DI, Lantos PL, eds), pp 197-262, 6th ed. Vol I: Arnold, London.
7. Vinters HV (1996) Brain parenchymal and microvascular amyloid in Alzheimer's disease. Brain Pathol 6: 179-195.
8. Scholz W (1938) Studien zur Pathologie der Hirngefäße II. Die drusige Entartung der Hirnarterien und -capillaren. (Eine Form seniler Gefäßerkrankung). Zeitschrift für die gesamte Neurologie und Psychiatrie 162: 694-715.
9. Weller RO, Massey AP, Newman TA et al. (1998) Cerebral amyloid angiopathy: β-amyloid accumulates in putative interstitial fluid drainage pathways in Alzheimer's disease. Am J Pathol 153: 725-733.
10. Shinkai Y, Yoshimura M, Ito Y et al. (1995) Amyloid β-proteins 1-40 and 1-42(43) in the soluble fraction of extra- and intracranial blood vessels. Ann Neurol 38: 421-428.
11. Ghetti B, Piccardo P, Frangione B et al. (1996) Prion protein amyloidosis. Brain Pathol 6: 127-145.

12. Mackie JB, Weiss MH, Miao W et al. (1998) Cerebrovascular accumulation and inceased blood-brain barrier permeability to circulating amyloid β peptide in aged squirrel monkey with cerebral amyloid angiopathy. J Neurochem 70: 210-215.
13. Harling-Berg CJ, Knopf PM, Merriam J et al. (1989) Role of the cervical lymph nodes in the systemic humoral immune response to human serum albumin microinfused into rat cerebrospinal fluid. J Neuroimmunol 25: 185-193.
14. Phillips MJ, Needham M, Weller RO (1997) Role of cervical lymph nodes in autoimmune encephalomyelitis in the Lewis rat. J Pathol 182: 457-464.
15. Kida S, Pantazis A, Weller RO (1993) CSF drains directly from the subarachnoid space into nasal lymphatics in the rat. Anatomy, histology and immunological significance. Neuropathol Appl Neurobiol 19: 480-488.
16. Nicholson C, Syková E (1998) Extracellular space structure revealed by diffusion analysis. Trends Neurosci 21: 207-215.
17. Zhang ET, Inman CBE, Weller RO (1990) Interrelationships of the pia mater and perivascular (Virchow Robin) spaces in the human cerebrum. J Anat 170: 111-123.
18. Storey E, Cappai R (1999) The amyloid precursor protein of Alzheimer's disease and the Aβ peptide. Neuropathol Appl Neurobiol 25: 81-97.
19. Lue LF, Kuo YM, Roher AE et al. (1999) Soluble amyloid β peptide concentration as a predictor of synaptic change in Alzheimer's disease. Am. J. Pathol. 155: 853-862.
20. McLean CA, Cherny RA, Fraser FW et al. (1999) Soluble pool of Aβ amyloid as a determinant of severity of neurodegeneration in Alzheimer's disease. Ann Neurol 46: 860-866.
21. Selkoe DJ (1999) Translating cell biology into therapeutic advances in Alzheimer's disease. Nature 399 (Suppl): A23-A31.
22. Roher AE (1993) β-amyloid (1-42) is a major component of cerebrovascular amyloid deposits: implications for the pathology of Alzheimer's disease. Proc Natl Acad Sci USA 90: 10836-10840.
23. Yamaguchi H, Yamazaki T, Lemere CA et al. (1992) β Amyloid is focally deposited within the outer basement membrane in the amyloid angiopathy of Alzheimer's disease. Am J Pathol 141: 249-259.
24. Calhoun ME, Burgermeister P, Phinney AL et al. (1999) Neuronal overexpression of mutant amyloid precursor protein results in prominent deposition of cerebrovascular amyloid. Proc Natl Acad Sci USA 96: 14088-14093.

TRANSGENIC MOUSE MODELS OF CEREBRAL AMYLOID ANGIOPATHY

[1]Martin C. Herzig, [1]David T. Winkler, [2]Lary C. Walker, and [1]Mathias Jucker
[1]Department of Neuropathology, Institute of Pathology, University of Basel, CH-4003 Basel, Switzerland; [2]Neuroscience Therapeutics, Pfizer Ann Arbor Laboratories, Ann Arbor, MI, USA

1. INTRODUCTION

The likelihood of amyloid deposition in the cerebral vasculature (cerebral amyloid angiopathy; CAA) increases with age[1-3]. Particularly in Alzheimer's disease (AD), CAA can be detected in up to 90% of the patients[4-6]. CAA of the ß-amyloid-type (Aß-CAA) affects primarily leptomeningeal and cortical vessels and is associated with degeneration of smooth muscle cells and endothelial cells, as well as blood-brain barrier damage. Severe Aß-CAA can cause fatal cerebral hemorrhage and increases the risk of cerebral microinfarcts[7-9]. Genetic variations that are causative or risk factors for Aß-CAA have been identified, such as mutations in the genes for ß-amyloid precursor protein (APP), presenilins 1 and 2, and possibly cystatin C, as well as polymorphisms in apolipoprotein E[3]. In normal aging and AD, Aß-CAA is associated with parenchymal Aß plaques. However, Aß-CAA can also occur in the absence of plaques, as evidenced in hereditary cerebral hemorrhage with amyloidosis-Dutch type (HCHWA-D)[10-12].

For many years, animal models of CAA have been limited to naturally occurring CAA in aged dogs and some aged primates[13]. These models have provided first insights into the pathogenesis of CAA. For example, Wisniewski and colleagues succeeded in culturing canine smooth muscle cells derived from amyloid-laden vessels and observed *in vitro* accumulation of intracellular Aß[14]. Walker and colleagues have demonstrated the predilection of CAA development in some monkey species. Aged squirrel

Neuropathology and Genetics of Dementia, Edited by Tolnay and Probst
Kluwer Academic/Plenum Publishers, New York 2001

monkeys (*Saimiri sciureus*) are prone to developing Aß deposition in the cerebral blood vessels, whereas aged rhesus monkeys (*Macaca mulatta*) develop preferentially parenchymal amyloid plaques[13]. Interestingly, both species have the same Aß sequence, however, a missense mutation (Leu68Met) in cystatin C at the Icelandic locus was identified in squirrel monkeys, which may explain their particular susceptibility to CAA[15].

2. APP TRANSGENIC MOUSE MODELS OF CAA

The first APP transgenic mouse to develop significant cerebral amyloidosis has been reported by Games and co-workers[16]. They used a platelet-derived growth factor (PDGF)-ß promoter to express a human APP (hAPP) minigene encoding the V717F mutation associated with familial AD. In these so-called PD-APP mice, overexpression of mutant hAPP is more than 10-fold higher than that of endogenous mouse APP. These mice progressively develop many of the pathological hallmarks of AD, including numerous extracellular, thioflavin S-positive Aß deposits, neuritic plaques, synaptic loss, astrocytosis and microgliosis, but no vascular amyloid was reported in these mice.

Subsequently, two other groups successfully created transgenic mice overexpressing hAPP bearing the Swedish double mutation (K670N, M671L). Hsiao and collaborators[17] made use of a prion protein (PrP) promoter to express the Swedish double mutation containing the APP_{695}-isoform in B6/SJL mice (Tg2576). They describe transgene-derived APP levels exceeding endogenous APP by 6-fold. Tg2576 mice develop cerebral amyloidosis beginning around 9 months of age[17]. Although initially not described, these mice also develop some CAA, but Aß deposition in the cerebral vasculature appears not to be a prominent feature in this model[3,18].

Sturchler-Pierrat and colleagues used the murine Thy-1 promoter to overexpress mutated human APP_{751} (APP23 mice)[19]. These mice show a 7-fold hAPP overexpression and amyloid plaques appear first at 6 months of age. The plaques are mostly congophilic and are accompanied by neuritic changes including dystrophic boutons. The plaques are surrounded by immunoreactivity for hyperphosphorylated tau, but, like all existing APP-transgenic models, the mice lack neurofibrillary tangles[19].

In contrast to the Tg2576 mouse model, CAA is a very prominent feature in aging APP23 mice, with striking similarities to that observed in human aging and AD. Amyloid deposition occurs preferentially in arterioles and capillaries, and within individual vessels shows a wide heterogeneity, ranging from a thin rim of amyloid in the vessel wall to vascular amyloid infiltrating the brain parenchyma. CAA in these mice is associated with perivascular neurodegeneration and microglial activation[20]. Most recently, a loss of vascular smooth muscle cells has been reported, leading to a weakening of the vessel wall that is eventually followed by rupture and hemorrhage[21].

3. MECHANISM OF CAA IN TRANSGENIC MICE

APP23 mice show high levels of Aß in CSF but no detectable levels in blood. Moreover, CAA is observed in areas without transgene expression, suggesting that axonal transport and diffusion rather than local production or blood uptake of Aß is key to the pathogenesis of CAA in APP23 mice[20]. To exclude a contribution of endogenous mouse Aß in the formation of CAA, APP23 mice were bred on an *APP*-null background. Such mice develop a similar degree of both plaques and CAA and provide further evidence that a neuronal source of Aß is sufficient to induce CAA and CAA-associated hemorrhage[20,21]. A recent study using the same approach has come to the same conclusion[22].

The most salient risk factors for CAA are advancing age[1,2,4], AD[4,5,23,24], and genetic influences[3,6]. Two genetic risk factors, namely apolipoprotein E (ApoE) and TGF-ß1, were tested by crossing hAPP transgenic mice with ApoE knockout and TGF-ß1 overexpressing mice[18,25]. Both crossings show no change in Aß production. However, in hAPP transgenic mice on an *ApoE*-null background, both amyloid plaques and CAA are drastically reduced, demonstrating that ApoE facilitates the formation of both parenchymal and cerebrovascular amyloid deposits[18]. On the other hand, cerebrovascular Aß deposition is accelerated selectively in hAPP/TGF-ß1 bigenic mice[25], indicating that CAA and extracellular amyloid deposition are independent processes, with aging and Aß as common risk factors.

4. CONCLUSION AND OUTLOOK

To optimize the experimental analysis of CAA, it will be important to produce mouse models that exclusively develop CAA. HCHWA-D patients harbor a point mutation within Aß at codon 693 of APP (E693Q); they develop severe CAA, and suffer fatal hemorrhagic strokes early in their fifties[12]. Patients with the Flemish APP mutation at the preceding codon (A692G) show a combined pathology of CAA and amyloid plaques[26,27], although most recent results suggest that most of the parenchymal amyloid is actually amyloid associated with vessels[28]. Transgenic mice with the Flemish and Dutch mutations have been generated. Such mice show some behavioral disturbances, however, they do not exhibit CAA up to 18 month of age[29].

Although the most common form of CAA is Aß-CAA, there are other proteins that have been linked to familial forms of CAA. Hereditary cerebral hemorrhage with amyloidosis-Icelandic type (HCHWA-I) is caused by a point mutation in the cystatin C gene, and leads to dementia and severe, early-onset hemorrhage that occurs at least two decades earlier than in HCHWA-D[30]. Patients with familial British dementia (FBD)[31] and familial Danish dementia (FDD)[32] develop severe CAA composed of ABri and ADan, respectively, but do not show significant bleedings. It is currently not clear whether the risk of hemorrhage is dictated by the type of amyloid protein that is deposited in the vascular wall, the amount of protein, the participation of cofactors such as pathological chaperones, or the anatomical distribution of the amyloid within the vessel and/or within certain brain regions.

In conclusion, although more comprehensive studies and further transgenic mouse models are necessary, the present results in transgenic mouse models have already improved the current understanding of the pathophysiology of CAA and have provided a basis for the development of effective interventions.

ACKNOWLEDGEMENTS

Our work on CAA is supported by the Swiss National Science Foundation, and the Fritz Thyssen Foundation (Cologne, Germany).

REFERENCES

1. Tomonaga M (1981) Cerebral amyloid angiopathy in the elderly. J Am Geriatr Soc 29: 151-157.
2. Vinters HV, Gilbert JJ (1983) Cerebral amyloid angiopathy: incidence and complications in the aging brain. II. The distribution of amyloid vascular changes. Stroke 14: 924-928.
3. Walker LC, Durham RA (1999) Cerebrovascular amyloidosis: experimental analysis in vitro and in vivo. Histol Histopathol 14: 827-837.
4. Esiri MM, Wilcock GK (1986) Cerebral amyloid angiopathy in dementia and old age. J Neurol Neurosurg Psychiatry 49: 1221-1226.
5. Yamada M, Tsukagoshi H, Otomo E et al. (1987) Cerebral amyloid angiopathy in the aged. J Neurol 234: 371-376.
6. Yamada M (2000) Cerebral amyloid angiopathy: an overview. Neuropathology 20: 8-22.

7. Vinters HV (1987) Cerebral amyloid angiopathy. A critical review. Stroke 18: 311-324.
8. Itoh Y, Yamada M, Hayakawa M et al. (1993) Cerebral amyloid angiopathy: a significant cause of cerebellar as well as lobar cerebral hemorrhage in the elderly. J Neurol Sci 116: 135-141.
9. Vonsattel JP, Myers RH, Hedley-Whyte ET et al. (1991) Cerebral amyloid angiopathy without and with cerebral hemorrhages: a comparative histological study. Ann Neurol 30: 637-649.
10. Wattendorff AR, Bots GT, Went LN et al. (1982) Familial cerebral amyloid angiopathy presenting as recurrent cerebral haemorrhage. J Neurol Sci 55: 121-135.
11. Haan J, Hardy JA, Roos RA (1991) Hereditary cerebral hemorrhage with amyloidosis--Dutch type: its importance for Alzheimer research. Trends Neurosci 14: 231-234.
12. Levy E, Carman MD, Fernandez-Madrid et al. (1990) Mutation of the Alzheimer's disease amyloid gene in hereditary cerebral hemorrhage, Dutch type. Science 248: 1124-1126.
13. Walker LC (1997) Animal models of cerebral beta-amyloid angiopathy. Brain Res Rev 25: 70-84.
14. Wisniewski HM, Frackowiak J, Mazur-Kolecka B (1995) In vitro production of beta-amyloid in smooth muscle cells isolated from amyloid angiopathy-affected vessels. Neurosci Lett 183: 120-123.
15. Wei LH, Walker LC, Levy E (1996) Cystatin C. Icelandic-like mutation in an animal model of cerebrovascular beta-amyloidosis. Stroke 27: 2080-2085.
16. Games D, Adams D, Alessandrini R, et al. (1995) Alzheimer-type neuropathology in transgenic mice overexpressing V717F beta-amyloid precursor protein. Nature 373: 523-527.
17. Hsiao K, Chapman P, et al. (1996) Correlative memory deficits, Abeta elevation, and amyloid plaques in transgenic mice. Science 274: 99-102.
18. Holtzman DM, Fagan AM, Mackey B et al. (2000) Apolipoprotein E facilitates neuritic and cerebrovascular plaque formation in an Alzheimer's disease model. Ann Neurol 47: 739-747.
19. Sturchler-Pierrat C, Abramowski D, Duke M et al. (1997) Two amyloid precursor protein transgenic mouse models with Alzheimer disease-like pathology. Proc Natl Acad Sci USA 94: 13287-13292.
20. Calhoun ME, Burgermeister P, Phinney AL et al. (1999) Neuronal overexpression of mutant amyloid precursor protein results in prominent deposition of cerebrovascular amyloid. Proc Natl Acad Sci U S A 96: 14088-14093.
21. Winkler DT, Bondolfi L, Herzig MC et al. (2000) Spontaneous hemorrhagic stroke in a mouse model of cerebral amyloid angiopathy. (submitted)
22. Van Dorpe J, Smeijers L, Dewachter I et al. (2000) Prominent cerebral amyloid angiopathy in transgenic mice overexpressing the London mutant of human APP in neurons. Am J Pathol (in press)
23. Mandybur TI (1975) The incidence of cerebral amyloid angiopathy in Alzheimer's disease. Neurology 25: 120-126.
24. Ellis RJ, Olichney JM, Thal LJ et al. (1996) Cerebral amyloid angiopathy in the brains of patients with Alzheimer's disease: the CERAD experience, Part XV. Neurology 46: 1592-1596.
25. Wyss-Coray T, Masliah E, Mallory M et al. (1997) Amyloidogenic role of cytokine TGF-beta1 in transgenic mice and in Alzheimer's disease. Nature 389: 603-606.

26. Hendriks L, van Duijn CM, Cras P et al. (1992) Presenile dementia and cerebral haemorrhage linked to a mutation at codon 692 of the beta-amyloid precursor protein gene. Nat Genet 1: 218-221.
27. Cras P, van Harskamp F, Hendriks L et al. (1998) Presenile Alzheimer dementia characterized by amyloid angiopathy and large amyloid core type senile plaques in the APP 692Ala-->Gly mutation. Acta Neuropathol 96: 253-260.
28. Kumar-Singh S, Ceuterick C, Lubke U et al. (2000) Pathology of Flemish APP692 Alzheimer's disease suggests that core containing plaques are angiocentric. Neurobiol Aging 21: S66.
29. Kumar-Singh S, Dewachter I, Moechars D et al. (2000) Behavioral disturbances without amyloid deposits in mice overexpressing human amyloid precursor protein with Flemish (A692G) or Dutch (E693Q) mutation. Neurobiol Dis 7: 9-22.
30. Levy E, Lopez-Otin C, Ghiso J et al. (1989) Stroke in Icelandic patients with hereditary amyloid angiopathy is related to a mutation in the cystatin C gene, an inhibitor of cysteine proteases. J Exp Med 169: 1771-1778.
31. Vidal R, Frangione B, Rostagno A et al. (1999) A stop-codon mutation in the BRI gene associated with familial British dementia. Nature 399: 776-781.
32. Vidal R, Revesz T, Rostagno A et al. (2000) A decamer duplication in the 3' region of the BRI gene originates an amyloid peptide that is associated with dementia in a Danish kindred. Proc Natl Acad Sci USA 97: 4920-4925.

ALPHA-SYNUCLEIN
Axonal transport, ligand interaction and neurodegeneration

[1]Poul Henning Jensen and [2]Wei Ping Gai
[1]Department of Medical Biochemistry, Bldg. 170, University of Aarhus, DK-8000 Aarhus C, Denmark, [2]Department of Physiology, Flinders Medical Centre, Bedford Park, SA 5042, Australia

1. INTRODUCTION

α-Synuclein is implicated in the pathogenesis of Parkinson's disease (PD). First, missense mutations in α-synuclein gene lead to familial autosomal dominant PD[1,2]. Second, abnormal fibrillar α-synuclein accumulate in Lewy bodies in degenerating nerve cells in PD and Dementia with Lewy bodies[3]. Third, transgenic expression of human α-synuclein in mice and fruitflies leads to the development of Lewy body-like inclusions and neurodegeneration[4,5]. The 140 amino acid α-synuclein is normally a presynaptic protein with no recognisable secondary structures[6]. In PD however, changes occur in its cellular localisation and secondary structure as the Lewy bodies contains α-synuclein filaments that resemble amyloid fibrils[7]. Accordingly, the α-synuclein filaments resemble the intracellular filaments formed in other neurodegenerative diseases by such as tau protein in Alzheimer's disease and huntingtin in Huntington's disease.

Neuropathology and Genetics of Dementia, Edited by Tolnay and Probst
Kluwer Academic/Plenum Publishers, New York 2001

2. AXONAL TRANSPORT

2.1 General aspects

Axonal transport delivers cellular components that are crucial for axonal and presynaptic function, e.g. organelles, enzymes, and lipids[8]. The axonal transport has traditionally been divided in individual rate components, fast anterograde and retrograde transport, slow component a and slow component b, that each represents groups of constituents moved by similar rates. Fast axonal transport (FC) represents the movement of vesicle populations at a velocity of 100-400 mm/day along microtubules driven by an expanding group of kinesin-like motorproteins. Slow component b (SCb) comprise a heterogeneous group of soluble proteins that moves at a rate of 2-4 mm/day. The molecular mechanisms responsible for this rate component is not clear. Slow component a (SCa) represents the slow movement (0.2-1 mm/day) of a more insoluble group of proteins that are considered to be incorporated in or tightly attached to the axonal cytoskeleton.

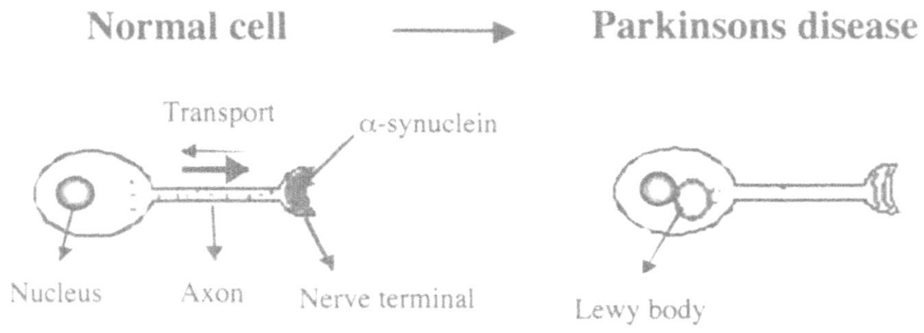

Figure 1. Relocalisation of α-synuclein from its normal presynaptic localisation during Parkinson's disease associated neurodegeneration. The Lewy body development may be dependent on abnormalities in the axonal transport machinery, increased fibrillogenic potential of the α-synuclein protein, increased concentration of fibrillogenic factors or combinations hereof.

2.2 Axonal α-synuclein transport

The axonal α-synuclein transport has been studied in the rat optic system[9]. In this model, retinal ganglion cells are metabolically pulse labeled where after axonally transported radioactive α-synuclein is immunoprecipitated from different segments of the about 2 cm long axonal pathway. The quantitative analysis revealed that α-synuclein is transported by all three rate components, with about 25% FC, 60% SCb and 15% SCa. This indicates that α-synuclein interacts with a range of axonal structures being the effectors of the different axonal transport modes. Axonal vesicles represent a carrier of FC components and α-synuclein do interact with brain vesicles. This interaction is mediated through determinants in the N-terminal 30 amino acid residues of α-synuclein but the vesicular binding partner is still unknown[10]. The interactions appears to require a high degree of structural stringency as the PD causing mutation Ala30Pro, but not the Ala53Thr mutation, inhibit the vesicle binding[10]. α-synuclein has the ability to interact with liposomes of acidic phospholipid[11], but this interaction does not likely reflect the binding to brain vesicles as it is independent of the N-terminal 30 amino acids (Lykkebo and Jensen, unpublished observation). Accordingly, a reduced axonal transport of α-synuclein(Ala30Pro) may contribute to an increased concentration of α-synuclein and subsequent Lewy body formation in the cell body of individuals carrying this mutation.

2.3 α-synuclein and microtubule-associated proteins

The components that govern the α-synuclein transport in SCa and SCb are unknown but the microtubule-associated protein tau is also transported by both SCa and SCb. Tau can bind to the C-terminal segment of α-synuclein[12]. A function of tau is to stabilise microtubules upon binding to these filamentous structures. Dysregulation of this binding may be of critical importance for the development of Alzheimer's disease and other neurodegenerative disorders where hyperphosphorylated tau is deposited as paired helical filaments in neurofibrillary tangles. A specific phosphorylation of Ser262, observed in paired helical filaments, inhibits the binding to and stabilisation of microtubules[13]. Binding of α-synuclein to tau makes Ser262 a better kinase substrate[12]. This indicates that α-synuclein can participate in the regulation of the axonal microtubule dynamics and may provide a link between perturbed axonal α-synuclein transport and more general properties of the axonal structure. A functional role for α-synuclein in nerve terminal function is suggested by a study of α-synuclein gene knockout mouse[14] and a nervecell culture study, where α-synuclein antisense suppression revealed a marked decrease in the distal pool of synaptic vesicles[15]. The structure of α-synuclein with separate domains for vesicle binding and protein ligands may contribute to vesicle homeostasis by linking specific vesicle pools to effectors of vesicle cycling. Moreover, disease associated dysfunctions in the axonal α-synuclein transport can hereby actively contribute to inefficient synaptic signalling.

The autosomal dominant inheritance pattern of the disease-causing α-synuclein mutations indicates a toxic gain of function by the mutant allele[1,2].

The abnormal accumulation of α-synuclein and Lewy body formation will expose constituents in the somatodendendritic compartment to high concentrations of α-synuclein. This may lead to interactions not normally occuring and result in a decrease function of vital factors due to a simple sequestration or a destabilisation. Microtubule associated protein 1B (MAP-1B) represents an example of a nervecell protein not normally in contact with significant amounts of α-synuclein. However, MAP-1B and α-synuclein colocalise in Lewy bodies and both proteins can be extracted from purified Lewy bodies[16]. A direct interaction is likely to occur as demonstrated by direct binding experiments between MAP-1B and both monomeric and filamentous α-synuclein[16].

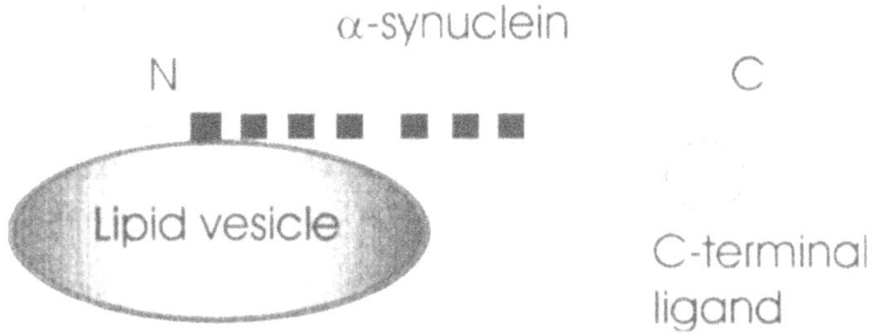

Figure 2. α-synuclein has the potential to function as a linking molecule by binding different ligands to separate domains. Brain vesicles bind to a site within the N-terminal 30 residues whereas C-terminally directed ligands, e.g. tau binds to the C-terminal 45 residues[10, 12].

3. OUTLOOK

A strong link has been established between neurodegeneration and abnormal α-synuclein metabolism. It is an attractive hypothesis to appoint the abnormal accumulation of large amount of filamentous α-synucleins in the somatodendritic compartment of neurons as the culprit, although Lewy bodies by some are considered as a protective arrangement whereby the cells may insulate toxic or malfunctioning proteins. Further unravelling of the pathobiological significance of Lewy bodies is expected with the development of transgenic animals recapitulating more aspects of Lewy body-associated neurodegeneration, the availability of large quantities of pure Lewy bodies from human brains[16, 17] and techniques for identifying α-synuclein filament binding proteins.

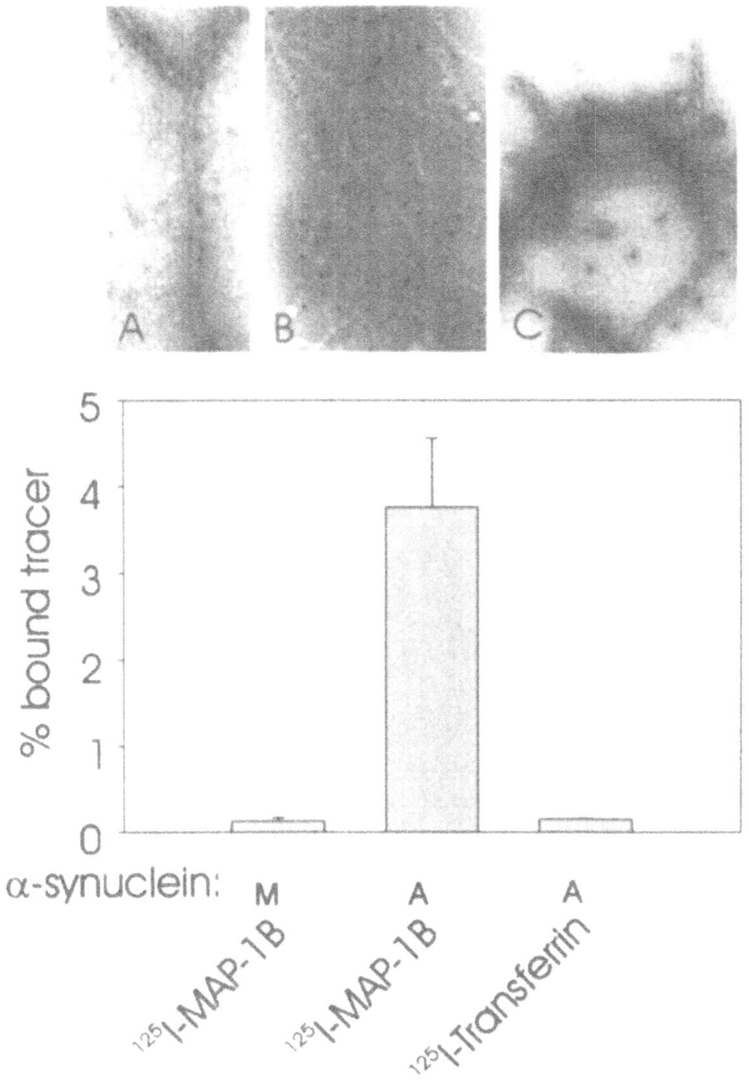

Figure 3. ^{125}I-MAP-1B bind to *in vitro* formed α-synuclein fibrils. Upper panel shows α-synuclein fibrils recovered from the pellet of the fibril binding assay, following labeling with antibody ASY-1 and secondary antibodies coupled to 15 nm gold particles. A, Straight and slightly curved fibrils. B, Dense meshes of α-synuclein fibrils. C, Negative control, processed as in A and B, but with the antibody ASY-1 replaced by preimmune rabbit IgG. The bar in C is 100 nm and also applies to A and B. Lower panel: Tracer binding to the pelleted material after incubation with monomeric (M) and fibrillar (A) α-synuclein. ^{125}I-transferrin, which does not bind α-synuclein, was used as negative control. The ordinate represents the percentage of the tracer recovered from the pellet. The bar represents the mean ± 1 S.D. of 3 independent experiments. The coefficient of variation between experiments was 0.15.

REFERENCES

1. Polymeropoulos MH, Lavedan C, Leroy E et al. (1997) Mutation in the α-synuclein gene identified in families with Parkinson's disease. Science 276: 2045-2047.
2. Krüger R, Kuhn W, Müller T et al. (1998) Ala30Pro mutation in the gene encoding α-synuclein in Parkinson's disease. Nat Genet 18: 106-108.
3. Spillantini MG, Schmidt ML, Lee VM-L et al. (1997) α-Synuclein in Lewy bodies. Nature 388: 839-840.
4. Masliah E, Rockenstein E, Veinbergs I et al. (2000) Dopaminergic loss and inclusion body formation in α-synuclein mice: implications for neurodegenerative disorders. Science 287: 1265-1269.
5. Feany MB, Bender WW (2000) A Drosophila model of Parkinson's disease. Nature 404: 394-398.
6. Weinreb PH, Zhen W, Poon AW et al. (1996) NACP, a protein implicated in Alzheimer's disease and learning, is natively unfolded. Biochemistry 35: 13709-13715.
7. Hashimoto M, Hsu LJ, Sisk A et al. (1998) Human recombinant NACP/alpha-synuclein is aggregated and fibrillated in vitro: relevance for Lewy body disease. Brain Res 799: 301-306.
8. Hirokawa N (1998) Kinesin and dynein superfamily proteins and the mechanism of organelle transport. Science 279: 519-526.
9. Jensen PH, Li JY, Dahlström A et al. (1999) Axonal transport of α-synuclein is mediated by all rate components. Eur J Neurosci 11: 3369-3376.
10. Jensen PH, Nielsen MS, Jakes R et al. (1998) Binding of α-synuclein to brain vesicles is abolished by familial Parkinson's disease mutation. J Biol Chem 273: 26292-26294.
11. Davidson WS, Jonas A, Clayton DF et al. (2000) Stabilization of alpha-synuclein secondary structure upon binding to synthetic membranes. J Biol Chem 273: 9443-9449.
12. Jensen PH, Hager H, Nielsen MS et al. (1999) α-synuclein binds to tau and stimulates the protein kinase A-catalysed tau phosphorylation of serine residues 262 and 356. J Biol Chem 274: 25481-25489.
13. Biernat J, Gustke N, Drewes G et al. (1993) Phosphorylation of Ser262 strongly reduces binding of tau to microtubules: distinction between PHF-like immunoreactivity and microtubule binding. Neuron 11: 153-166.
14. Abeliovich A, Schmitz Y, Farinas I et al. (2000) Mice lacking α-synuclein display functional deficits in the nigro-striatal dopamine system. Neuron 25: 239-252.
15. Murphy DD, Ructer SM, Trojanowski JQ et al. (2000) Synucleins are developmentally expressed and α-synuclein regulates the size of presynaptic vesicular pool in primary hippocampal neurons. J Neurosci 20: 3214-3220.
16. Jensen PH, Islam K, Kenney J et al. (2000) Microtubule-associated protein 1B is a component of cortical Lewy bodies and binds α-synuclein filaments. J Biol Chem 275: 21500-21507.
17. Gai WP, Power JHT, Blumbergs PC et al. (1999) Alpha-synuclein immunoisolation of glial inclusions from multiple system atrophy brain tissue reveals multiprotein components. J Neurochem 73: 2093-2100.

DEMENTIA WITH LEWY BODIES

Paul Ince
Neuropathology, Division of Genomic Medicine, Sheffield University Medical School, Sheffield, U.K.

1. INTRODUCTION

An increasing variety of neurological syndromes and brain disorders have been associated with Lewy body pathology in various parts of the nervous system[1]. The major dementia syndromes in which Lewy bodies are frequently identified are Dementia with Lewy bodies (DLB), dementia arising in Parkinson's disease, and Alzheimer's disease. Within each of these clinico-pathological umbrellas many patients can be allocated to the core syndromes based on a combination of clinical features and the pathological changes demonstrable at autopsy. However, there is a substantial, and currently unquantified, overlap between them such that some authors have proposed a continuum of Lewy body and Alzheimer-type pathologies[2]. The paradigms of Parkinson's disease and Alzheimer's disease have been absorbed into clinical practice, pathological interpretation at autopsy, and models of pathogenesis, to such an extent that DLB has been difficult to assimilate. DLB is currently defined in terms of a clinical syndrome and prospective pathological studies suggest that this syndrome predicts the presence of Lewy body pathology at least as well as a diagnosis of AD can predict the pathology of that disorder[3]. Similar considerations apply to Parkinson's disease so that for each of the three syndromes the currently accepted clinical criteria for their diagnosis is likely to be verified at autopsy in at least 4 out of 5 cases. The clinical features of DLB are defined in the published consensus criteria from an International Workshop held in 1995[4]. Many studies published subsequently have turned the concept of DLB

Neuropathology and Genetics of Dementia, Edited by Tolnay and Probst
Kluwer Academic/Plenum Publishers, New York 2001

around and have attempted, successfully, to prove that the clinical criteria do not reliably predict which cases will have Lewy body pathology at autopsy. This is because it is increasingly apparent that Lewy bodies are not uncommonly present in some brain areas in otherwise typical AD patients[5-6]. If the disorder were called, for example, "Kosaka's disease"[7-10], or some other eponymous name, then the inconsistent relationship between the clinical disorder and pathological findings would probably be regarded as no more problematic than is the case for AD and PD. At present there are major clinical reasons for adopting the existing terminology and diagnostic criteria. DLB in routine practice is predominantly an issue in the differential diagnosis of Alzheimer's disease and it is clear from a number of studies that patients with DLB are: 1) likely to die earlier, 2) show serious and life threatening reactions to major tranquilliser therapy, and 3) respond positively to anticholinesterase therapy, compared with patients whose dementia is associated with Alzheimer-type pathology alone[11-15].

The present review is chiefly concerned with the relationship between AD, PD and DLB at a clinicopathological level. The underlying pathogenetic relationship between the disorders is not clear but must involve as yet undetermined interactions between the three major pathways of protein dysmetabolism (tau, amyloid precursor protein and α-synuclein) that characterise these disorders.

2. PATHOLOGICAL MARKERS OF LEWY BODY PATHOLOGY

Lewy bodies were originally described in the pigmented monoaminergic and catecholaminergic neurons of the diencephalon and upper brainstem. In these neurons Lewy bodies stand out in conventionally stained sections as rounded intracytoplasmic inclusions characterised by a dense eosinophilic core with a surrounding pale halo. Studies in PD have extended the range of anatomical sites for Lewy body formation very considerably (Table 1). Pathological examination of PD cases also reveals a different form of Lewy body pathology which is present in the proximal dendritic region of some neuronal populations, notably the dorsal efferent nucleus of the vagus nerve, known as "Lewy neurites". The chief reason for the relatively recent description of DLB as an entity seems to be a consequence of the much less obvious morphology of Lewy bodies in cortical neurons using conventional stains. Cortical Lewy bodies (cLB) lack the intense eosinophilia and peripheral lucency of their subcortical counterpart and were undoubtedly overlooked in most pathological studies of Alzheimer's disease cohorts published before 1989. Initial attempts to improve the ease of detection of cLB showed that immunocytochemistry for ubiquitin is of value[16,17]. However, the other lesion which is commonly present in aged demented patients, the neurofibrillary tangle of Alzheimer-type pathology, is also ubiquitinated. There is therefore a need to exclude the presence of tangles in small globular neurons, by demonstrating negative tau immunocytochemistry, before accepting lesions as Lewy bodies using this approach.

Ventral tegmental area
Pedunculopontine tegmental nucleus
Locus ceruleus
Raphe subdivisions
Substantia nigra
Nucleus basalis of Meynert
Limbic cortex, especially amygdala
Temporal neocortex
Spinal autonimic ganglia
Myenteric sympathetic/parasympathetic neurons

Table 1. Anatomical regions regularly containing Lewy bodies in classical idiopathic Lewy body Parkinson's disease

Most recently it has been demonstrated that a major component, possibly the core filamentous constituent, of Lewy bodies is the protein α-synuclein[18-20]. This provides an ideal antigen for the specific detection of Lewy bodies and a number of excellent antibody reagents are available commercially and through research laboratories. The availability of α-synuclein immunocytochemistry is rapidly changing accepted views of the pathology of DLB, PD and AD. In particular it is much more sensitive than either conventional stains or ubiquitin immunocytochemistry. The pathological section of the DLB Consensus statement published in 1996 would be different if formulated today[4].

3. THE PATHOLOGY OF DLB

3.1 α-synuclein pathology

The only absolute pathological requirement for a diagnosis of DLB is the presence of Lewy bodies. However, while many demented patients with severe Alzheimer-type pathologies have some Lewy bodies in their brain[6], and therefore undoubtedly have a dementia with Lewy bodies, it is only if they have clinical features consistent with the Consensus Guidelines that they should be given the diagnosis of "Dementia with Lewy bodies". Most of the earlier literature on DLB is concerned with the distribution of Lewy bodies in regions classically associated with PD (e.g. substantia nigra, locus ceruleus, dorsal vagus nucleus) compared with the variable involvement of cortical regions. The advent of α-synuclein immunocytochemistry is now widening such studies to many other brain regions (e.g. the basal ganglia and thalamus) and emphasises the presence of previously undetected α-synuclein pathology in the form of diffuse neuritic lesions in neuronal processes (apparently both dendritic and axonal)[20].

The two principal pathological approaches to evaluating Lewy body pathology are semi-quantitative and can be used together. Subcortical involvement of the substantia nigra, locus ceruleus and vagus nucleus are included in the CERAD protocol[21] such that the frequency of Lewy bodies and the extent of associated neuronal loss are graded as absent, mild, moderate or severe. Patients with a clinical presentation of Parkinson's disease will usually have a loss of pigmented nigral neurons of at least 70% and there is usually significant depletion of the locus ceruleus. In contrast patients with DLB have much more variable pathology affecting the substantia nigra. In the original studies published this was frequently no more than 30%[15,22], and in some cases was minimal, reflecting the variable and mild parkinsonism which characterises DLB.

Cortical Lewy body pathology can be assessed using the protocol proposed by the DLB International Workshop[4]. This uses three neocortical (frontal, temporal and parietal) and two limbic cortical (transentorhinal and anterior cingulate) areas and was 'titrated' on the basis of experience with either haematoxylin and eosin staining or ubiquitin immunocytochemistry, since it predates α-synuclein immunocytochemistry. For each area a score of 0, 1, or 2 is allocated corresponding to the presence of 0, less than 5, or 5+ Lewy bodies in each region respectively. These scores are added together and the resulting total (between 0-10) subdivided into: 0-2 "brainstem predominant", 3-6 "limbic/transitional", and 7-10 "neocortical". There is no requirement that cortical Lewy body pathology reach any particular threshold before a diagnosis of DLB can be assigned. In the presence of the characteristic clinical syndrome a patient can be assigned as DLB even if Lewy bodies are confined to the brainstem or other regions not included in the Consensus cLB scoring method. This protocol for cLB assessment needs to be revised for several reasons related to the availability of α-synuclein immunocytochemistry. Firstly α-synuclein reveals far more pathology than the previously used techniques, so that use of the threshold of 5 lesions per area is likely to 'inflate' the cLB score. This remains a theoretical assumption because as yet there has been no publication which directly shows the effect on the cLB score of using α-synuclein versus ubiquitin on the same group of cases. Secondly, the Consensus cLB protocol does not take into account the amygdala which is an area that appears to be affected in a higher proportion of cases than most other cortical regions[6,23]. Thirdly, the relative significance of neuritic α-synuclein pathology, compared with Lewy bodies themselves, has not been established in terms of their role as the pathological substrate for the clinical manifestations of the disorder. This problem needs to be addressed in future studies both in terms of the quantitative burden of neuritic pathology in brain regions susceptible to Lewy body formation and also in the many brain regions which have not yet been systematically assessed in previous work.

3.2 ALZHEIMER-TYPE PATHOLOGY IN DLB

The extent and distribution of senile plaques and neurofibrillary tangles in DLB is highly variable[15,24-38]. This problem has been approached in the literature from two perspectives: Firstly there are studies in which the starting point is the selection of cases, either prospectively or by retrospective case-note review, using the Consensus diagnostic criteria for DLB, secondly there are studies which take an autopsy cohort of demented patients and seek to define the clinical and pathological features associated with the presence of Lewy bodies.

Starting from a clinical diagnosis compatible with DLB, there may be a wide range of severity of Alzheimer-type pathology. In general senile plaques are present in many cases at a level which is indistinguishable from AD[15,27,30]. However, even when the plaque count is high the DLB cases have mild or insignificant neocortical Alzheimer-type tau pathology which is reflected in a paucity of neocortical neurofibrillary tangles[15,30], absence of any diffuse increase in insoluble tau species[39-40], and a lack of tau immunoreactivity in the neuritic component of senile plaques[39-41]. These distinctions do not apply to limbic and archicortical regions which do contain neurofibrillary tangles and diffuse neuritic changes[33,42,43]. However, a proportion, possibly the minority, of elderly DLB patients have minimal Alzheimer-type pathology in any brain region. This range of Alzheimer-type pathology has given rise to the concept of two pathological subgroups of DLB patients[32]. The first of these, characterised by significant neocortical βA4 plaque pathology and medial temporal lobe tau pathology has been called the "common form" of DLB, indicating that these patients are in the majority. The second group with no more than minor changes of Alzheimer-type pathology are known as the "pure form" of DLB. This view has both uses and limitations. At present there are no agreed clinical characteristics which allow the separation of these groups prospectively. This is very interesting because it would seem to indicate that the Lewy body pathology is likely to be the major determinant of the functional changes in the brain rather than plaque amyloidosis. Work prior to the era of α-synuclein research emphasised the relatively low frequency of Lewy bodies in DLB (approximately 1000x less abundant / unit area of cortex), compared to neurofibrillary tangles in AD, despite an equivalent severity of dementia. The realisation, through α-synuclein immunocytochemistry, that DLB cases have widespread neuritic pathology[19,20] will have a major impact on the development of clinicopathological correlations for DLB and other Lewy body disorders. Indeed there is a strong argument for regarding these disorders as the "α-synucleinopathies".

4. LEWY BODY PATHOLOGY IN AD

There is an accumulating literature which criticises the concept of "Dementia with Lewy bodies" on the basis that Lewy body pathology is present in many cases of Alzheimer's disease. The reported prevalence of cases with Lewy bodies who meet diagnostic criteria for Alzheimer's disease varies from 7-71%[6,22,29,30,44-48]. The reasons for this enormous variation are not difficult to enumerate: The criteria used for the diagnosis of AD have evolved from reliance on the burden of cortical amyloidosis alone[49], through a method relying on cortical neuritic plaques (CERAD)[21] to the latest NIA-Reagan Institute guidelines which propose that the correlation between pathology and dementia is highest when there are both abundant cortical plaques and tangles[50]. Different research groups have access to autopsy cases through various routes, most of these series introduce biases related to clinical practice, voluntary enrolment in research cohorts etc. For example, one of the studies with the lowest proportion[22] sought to deliberately recruit patients with few atypical clinical features. None of these reports describe true, population-based cohorts. Secondly the method used for the detection of Lewy body pathology strongly influences the likelihood of detecting them. Most predate the use of α-synuclein immunocytochemistry which is the most sensitive method. The study of Hamilton[6] illustrates the need to sample the most commonly affected anatomical regions and draws particular attention to the amygdala which many previous studies have neglected. This most recent study used α-synuclein immunocytochemistry and demonstrated Lewy bodies in 61% of a group of 145 patients fulfilling the CERAD criteria for AD. However, because DLB patients frequently have high plaque counts many fulfil the pathological criteria for AD using the CERAD protocol. The proportion of such cases in a recent prospective study of patients meeting the Consensus Clinical criteria for DLB was >50%[3]. If the diagnosis of AD is made using criteria that require the presence of neocortical tangles then the prevalence of Lewy body pathology was 57% in the study of Hamilton[6] which is still a very considerable proportion. There are two particular points of interest from that study. Firstly it rightly emphasises the existence of many demented patients with Lewy bodies whose major clinical and pathological features are those of AD. I have used the term "AD with incidental Lewy bodies" to describe these patients[22] and they have been neglected in the earlier accounts of DLB[15,29]. There is now clear evidence that up to 60% of patients with familial early onset AD, due to either amyloid precursor protein or presenilin gene mutations, have Lewy body pathology in addition to the severe Alzheimer-type pathology that characterises their disorder[5,,51]. It has also been shown that elderly patients with Down's syndrome who develop severe Alzheimer-type pathology also accumulate Lewy bodies in regions such as the amygdala[52]. These data emphasise that it is not the presence of Lewy bodies alone which determines the clinical syndrome with which the patients present. The second finding in the study of Hamilton[6] with considerable potential impact on research in DLB is the prominence of α-synuclein pathology in the amygdala. The Pathology group of the International Workshop[4] adopted the term "brainstem predominant" for the group of patients with least abundant Lewy body pathology. At that time there was no systematic study of amygdala pathology in DLB and the prevailing view was that if there was cortical involvement then there would invariably be some degree of Lewy body pathology in the pigmented monoaminergic and catecholaminergic brainstem nuclei. This concept was strongly influenced by experience of Parkinson's disease and the notion that DLB cases usually had some mild parkinsonian features. This recent work now shows that α-synuclein pathology in the amygdala can be marked in the absence of apparent midbrain/brainstem involvement. While it is possible to dispute how meticulously the nigra should be examined before declaring it quite free of α-synuclein pathology, it is clear that future studies need to examine cases from the perspective that the amygdala may be the "centre of gravity" of Lewy body formation. A problem in interpreting Hamilton's study is the lack of information as to whether the DLB diagnostic guidelines were used in addition to the NINDS-ADRDA criteria for AD[53] since many DLB cases will also meet criteria for probable AD[3]. Despite this caveat there is good evidence that some AD patients will have quite extensive cortical α-synuclein pathology while others will have it in a distribution limited to the amygdala and adjacent entorhinal cortex.

5. WHICH PATHOLOGY IS FUNCTIONALLY DOMINANT IN DLB AND AD?

Because the pathological correlates of DLB and AD involve major overlapping contributions of senile plaque formation and variable α-synuclein and tau pathologies it is of interest to consider to what extent any of these processes can be regarded as the substrate for the particular clinical profile which each syndrome displays.

Amyloid plaque formation is a poor candidate. Cases of DLB often have an equivalent burden of senile plaques, in the same cortical regions as classical AD patients, despite presenting with a very different clinical syndrome. It is also well recognised that intellectually normal elderly people can also have an intense level of cortical amyloid deposition[54].

In a similar way the presence of α-synuclein immunoreactive inclusions and neurites occurs frequently in AD patients in the absence of the clinical features of AD. In a recent study longitudinally assessed patients with dementia were followed up in a research clinic until their death with subsequent autopsy examinations[24]. The pathological data was used independently of the clinical data to group the cases into "pure AD", pure DLB, and "mixed AD + DLB". The clinical data were then analysed to look for discriminating variables that characterised these groups. The "pure DLB" pathology group had most of the typical clinical features that are used to make the clinical diagnosis of DLB. By contrast the "AD+DLB" group were virtually indistinguishable from the "pure AD" pathology group. This latter group were defined on the basis of at least minor neocortical neurofibrillary tangle pathology. Therefore this study, and the experience of others previously published, suggests that when neocortical tau pathology is significant the clinical dementia phenotype will conform to conventional AD whether or not there is significant α-synuclein pathology. In contrast cerebral α-synuclein pathology on its own is likely to produce a clinical disorder within the spectrum of the Lewy body disorders[1,55,56] whether or not it is associated with both significant amyloid plaque formation and moderate medial temporal lobe tau pathology.

REFERENCES

1. Ince PG, Perry EK, Morris CM (1998) Dementia with Lewy bodies. A distinct non-Alzheimer dementia syndrome? Brain Pathol 8: 299-324.
2. Perl DP, Olanow CW, Calne D (1998) Alzheimer's disease and Parkinson's disease: distinct entities or extremes of a spectrum of neurodegeneration. Ann Neurol 44 (Suppl 1): S19-S31.

3. McKeith I, Ballard C, Perry R et al. (2000) Prospective validation of Consensus criteria for the diagnosis of Dementia with Lewy bodies. Neurology 54: 1050-1058.
4. McKeith IG, Galasko D, Kosaka K et al. (1996) Consensus guidelines for the clinical and pathological diagnosis of dementia with Lewy bodies (DLB): report of the consortium on DLB International Workshop. Neurology 47: 1113-1124.
5. Lantos PL, Ovenstone IM, Johnson J et al. (1994) Lewy bodies in the brains of two members of a family with the 717 (Val to Ile) mutation of the amyloid precursor protein gene. Neurosci Lett 172: 77-79.
6. Hamilton RL (2000) Lewy bodies in Alzheimer's disease: a neuropathological review of 145 cases using α-synuclein immunohistochemistry. Brain Pathol 10: 378-384.
7. Kosaka K (1978) Lewy bodies in the cerebral cortex. Report of three cases. Acta Neuropathol 42: 127-134.
8. Kosaka K, Mehraein P (1979) Dementia-parkinsonism with numerous Lewy bodies and senile plaques in cerebral cortex. Arch Psychiatr Nervenkr 226: 241-250.
9. Kosaka K, Matsushita M, Oyanagi S et al. (1980) A clinicopathological study of the "Lewy Body Disease". Pschiat Neurol Japon 82: 292-311.
10. Kosaka K, Yoshimura M, Ikeda K et al. (1984) Diffuse type of Lewy body disease: progressive dementia with abundant cortical Lewt bodies and senile changes of varying degree - A new disease? Clin Neuropathol 3: 185-192.
11. Cummings J (1997) Changes in neuropsychiatric symptoms as outcome measures in clinical trials with cholinergic therapies for Alzheimer's disease. Alz Dis Assoc Disord 11: 51-59.
12. Hutchinson M, Fazzini E (1996) Cholinesterase inhibition in Parkinson's disease. J Neurol Neurosurg Psychiatry 61: 324-325.
13. McKeith IG, Fairbairn A, Perry RH et al. (1992) Neuroleptic sensitivity in patients with senile dementia of Lewy body type. BMJl 305: 673-678.
14. Olichney JM, Galasko D, Salmon DP et al. (1998) Cognitive decline is faster in Lewy body variant than in Alzheimer's disease. Neurology 51:351-357.
15. Perry RH, Irving D, Blessed G et al. (1990) Senile dementia of Lewy body type: a clinically and neuropatholgically distinct form of Lewy body dementia in the elderly. J Neurol Sci 95: 119-139.
16. Lennox G, Lowe J, Morrell K et al. (1989) Anti-ubiquitin immunocytochemistry is more sensitive than conventional techniques in the detection of diffuse Lewy body disease. J Neurol Neurosurg Psychiatry 52: 67-71.
17. Lowe J, Blanchard A, Morrell K et al. (1988) Ubiquitin is a common factor in intermediate filament inclusion bodies of diverse type in man, including those of Parkinson's disease, Pick's disease and Alzheimer's disease as well as Rosenthal fibres in cerebellar astrocytomas, cytoplasmic bodies in muscle and Mallories bodies in alcoholic liver disease. J Pathol 155: 9-15.
18. Polymeropoulos MH, Lavedan C, Leroy E et al. (1997) Mutation in the α-synuclein gene identified in families with Parkinson's disease. Science 276: 2045-2047.
19. Spillantini MG, Schmidt ML, Lee VM-Y et al. (1997) α–Synuclein in Lewy bodies. Nature 388: 839-840.
20. Spillantini MG, Crowther RA, Jakes R et al. (1998) α-synuclein in filamentous inclusions of Lewy bodies from Parkinson's disease and Dementia with Lewy bodies. Proc Natl Acad Sci USA 95: 6469-6473.
21. Mirra SS, Heyman A, McKeel D et al. (1991) The consortium to establish a registry of Alzheimer's disease (CERAD) Part II. Standardization of the neuropathologic assessment of Alzheimer's disease. Neurology 41: 479-486.

22. Ince PG, McArthur FK, Bjertness E et al. (1995) Neuropathological diagnoses in elderly patients in Oslo: Alzheimer's disease, Lewy body disease and vascular lesions. Dementia 6: 162-168.
23. Braak H, Braak E, Yilmazer D et al. (1994) Amygdala pathology in Parkinson's disease. Acta Neuropathol 88: 493-500.
24. Del Ser T, Hachinski V, Merskey H et al. (2000) Clinical and pathological features of two groups of dementia with Lewy bodies: effect of co-existing Alzheimer-type lesion load. Alz Dis Assoc Disorders: in press.
25. Dickson DW, Davies P, Mayeux R et al. (1987) Diffuse Lewy body disease: Neuropathological and biochemical studies of six patients. Acta Neuropathol 75: 8-15.
26. Galasko D, Katzman R, Salmon DP et al. (1996) Clinical features and neuropathological findings in Lewy body dementias. Brain and Cognition 31: 166-175.
27. Gentleman SM, Williams B, Royston MC et al. (1992) Quantification of ßA4 protein deposition in the medial temporal lobe: A comparison of Alzheimer's disease and senile dementia of Lewy body type. Neurosci Lett 142: 9-12.
28. Hansen LA, Masliah E, Terry RD et al. (1989). A neuropathological subset of Alzheimer's disease with concomitant Lewy body disease and spongiform change. Acta Neuropathol 78: 194-201.
29. Hansen L, Salmon D, Galasko D et al. (1990) The Lewy body variant of Alzheimer's disease: a clinical and pathological entity. Neurology 40: 1-8.
30. Hansen LA, Masliah E, Galasko D et al. (1993) Plaque-only Alzheimer's disease is usually the Lewy body variant and vice versa. J Neuropathol Expl Neurol 52: 648-654.
31. Hansen LA, Samuel W (1997) Criteria for Alzheimer's disease and the nosology of dementia with Lewy bodies. Neurology 48: 126-132.
32. Hansen LA (1997) The Lewy body variant of Alzheimer disease. J Neural Transm 51[Suppl]: 83-93.
33. Ince P, Irving D, MacArthur F et al. (1991) Quantitative neuropathology in the hippocampus: comparison of senile dementia of Alzheimer type, senile dementia of Lewy body type, Parkinson's disease and non-demented elderly control patients. J Neurol Sci 106: 142-152.
34. Jellinger KA, Bancher C (1996) Dementia with Lewy bodies: relationships to Parkinson's and Alzheimer's diseases. In: Dementia with Lewy Bodies (Perry RH, McKeith IG, Perry EK, eds), pp 268-286: Cambridge University Press, Cambridge.
35. Kosaka K (1993) Dementia and neuropathology in Lewy body disease. Adv Neurol 60: 456-463.
36. Samuel W, Crowder R, Hofstetter CR et al. (1997) Neuritic plaques in the Lewy body variant of Alzheimer's disease lack paired helical filaments. Neurosci Lett 223: 73-76.
37. Samuel W, Alford M, Hofstetter CR et al. (1997) Dementia with Lewy bodies versus pure Alzheimer's disease: differences in cognition, neuropathology, cholinergic dysfunction, and synaptic density. J Neuropathol Exp Neurol 56: 499-508.
38. Sima AAF, Clark AW, Sternberger NA et al. (1986) Lewy body dementia without Alzheimer changes. Can J Neurol Scis 13(Suppl)): 490-497.
39. Harrington CR, Perry RH, Perry EK et al. (1994) Senile dementia of Lewy body type and Alzheimer type are biochemically distinct in terms of paired helical filaments and hyperphosphorylated tau protein. Dementia 5: 216-228.
40. Strong C, Anderton BH, Perry RH et al. (1995) Abnormally phosphorylated tau protein in senile dementia of Lewy body type and Alzheimer's disease: evidence that the disorders are distinct. Alz Dis Assoc Disorders 9: 218-222.

41. Dickson DW, Crystal H, Mattiace LA et al. (1989) Diffuse Lewy body disease: light and electron microscopic immunocytochemistry of senile plaques. Acta Neuropathol 78: 572-584.
42. Braak H, Braak E, Yilmazer D et al. (1996) Pattern of brain destruction in Parkinson's disease and Alzheimer's disease. J Neural Transm 103: 455-490.
43. Dickson DW, Schmidt ML, Lee VM et al. (1994) Immunoreactivity profile of hippocampal CA2/3 neurites in diffuse Lewy body disease. Acta Neuropathol 87: 269-276.
44. Forstl H, Burns A, Luthert P et al. (1993) The Lewy body variant of Alzheimer's disease: clinical and pathological findings. Br J Psychiat 162: 385-392.
45. Galasko D, Hansen LA, Katzman R et al. (1994) Clinical-neuropathological correlation's in Alzheimer's disease and related disorders. Arch Neurol 51:888-895.
46. Heyman A, Fillenbaum GG, Gearing M et al. (1999) Comparison of Lewy body variant of Alzheimer's disease with pure Alzheimer's disease: Consortium to Establish a Registry for Alzheimer's disease, Part XIX. Neurology 52: 1839-1844.
47. Kazee AM, Han LY (1995) Cortical Lewy bodies in Alzheimer's disease. Arch Path Lab Med 119: 448-453.
48. Victoroff J, Mack WJ, Leyness SA et al. (1995) Multicentre clinicopathological correlation in dementia. Am J Psychiat 152: 1476-1484.
49. Khachaturian ZS (1985) The diagnosis of Alzheimer's disease. Arch Neurol 42:1097-1105.
50. The National Institute on Aging and Reagan Institute Working Group on Diagnositic Criteria for the Neuropathological Assessment of Alzheimer's disease (1997) Consensus recommendations for the postmortem diagnosis of Alzheimer's disease. Neurobiol Aging 18 (Suppl 4): S1-27.
51. Lippa CF, Fujiwara H, Mann DM et al. (1998) Lewy bodies contain altered alpha-synuclein in brains of many familial Alzheimer's disease patients with mutations in presenilin and amyloid precursor protein genes. Am J Path 153: 1365-1370.
52. Lippa CF, Scmidt ML, Lee VM et al. (1994) Antibodies to alpha-synuclein detect Lewy bodies in many Down's syndrome brains with Alzheimer's disease. Ann Neurol 45: 353-357.
53. McKhann G, Drachman D, Folstein M et al. (1984) Clinical diagnosis of Alzheimer's Disease: report of the NINCDS-ADRDA Workgroup under the auspices of the Department of Health and Human Services Task Force on Alzheimer's Disease. Neurology 34: 939-944.
54. Tomlinson BE, Blessed G, Roth M (1968) Observations on the brains of non-demented old people. J Neurol Sci 7: 331-356.
55. Kosaka K, Iseki E (1996) Diffuse Lewy body disease within the spectrum of Lewy body disease. In: Dementia with Lewy Bodies (Perry RH, McKeith IG, Perry EK, eds), pp 238-247: Cambridge University Press, Cambridge.
56. Ince P, Perry R, Perry E (2000) Pathology of dementia with Lewy bodies. In: Dementia (O'Brien J, Burns A, eds), pp 699-717, 2nd ed: Arnold, London.

MOUSE MODELS OF α-SYNUCLEINOPATHY AND LEWY PATHOLOGY

α-synuclein expression in transgenic mice

[1]Samuel Barbieri, [1]Katja Hofele, [1]Karl-Heinz Wiederhold, [2]Alphonse Probst, [2]Claudia Mistl, [1]Simone Danner, [1]Sabine Kauffmann, [1]Bernd Sommer, [1]Willibrordus Spooren, [2]Markus Tolnay, [1]Graeme Bilbe, and [1]Herman van der Putten

[1]Nervous System Research, Novartis Pharma Inc., CH 4002 Basel, Switzerland, [2]Institute for Pathology, Schönbeinstrasse 40, CH 4003 Basel, Switzerland.

1. INTRODUCTION

Idiopathic Parkinson's Disease (PD), dementia with Lewy bodies (DLB), and a Lewy Body variant of Alzheimer's Disease (LBVAD) are characterized pathologically by proteinaceous inclusions in neurons commonly referred to as Lewy pathology in *post mortem* brain tissue samples[1-6]. The inclusions occur in the dystrophic (Lewy) neurites that constitute an important part of the pathology of PD and DLB[4-6], in neuronal perikarya (Lewy bodies), and occasionally extracellularly[11].

α-Synuclein (αSN) is the major constituent of Lewy bodies and Lewy neurites[2-10,12,13]. To lesser and varying degrees, these inclusions contain ubiquitin[14], neurofilament proteins[1-3], ubiquitin C-terminal hydrolase[15], proteasomal subunits[16] and a variety of other antigenic determinants[17]. αSN immunoreactivity is the most sensitive and reliable diagnostic for Lewy-type pathology[1-10].

Lewy pathology in neurons appears to be central and may contribute mechanistically to their dysfunction and degeneration in disease. Altogether, the distribution of αSN containing inclusions in disorders with Lewy pathology, the discovery of two mutations in the αSN gene linked to early-onset familial PD[18,19], and the ability of the protein to self-aggregate[20-24], support a central role for αSN in the pathophysiology of diseases with Lewy pathology. Recent findings *in vivo* in transgenic flies[25] and transgenic mice[26-

Neuropathology and Genetics of Dementia, Edited by Tolnay and Probst
Kluwer Academic/Plenum Publishers, New York 2001

[28] provide additional support for this notion and novel perspectives to model in animals the pathophysiology of diseases with α-synucleinopathy.

2. CLINICAL-PATHOLOGICAL CONSIDERATIONS.

When trying to develop animal models for diseases with Lewy pathology, it is important to keep in mind that in these diseases, damage generally occurs to multiple systems. It is well known that the major clinical symptoms of PD (tremor, rigidity, akinesia, and the response of patients to dopamine replacement therapy) are related to the loss of dopaminergic neurons in the *substantia nigra* (SN). But in addition to the damage in this brain region and the loss of dopaminergic neurons, serotonergic, cholinergic and other catecholaminergic neurons and nuclei are often affected. As a result, complex biochemical and pathological changes may occur and these could well represent the basis for the variable clinical picture associated with PD and DLB with respect to e.g. motor, vegetative, behavioural and cognitive dysfunction. It seems far fetched to try and mimic all of the complex changes that might occur in a single animal model while still trying to understand the consequences of damage occurring in individual brain nuclei. Rather, it seems likely that more precise information will be obtained from animal models with different and neuronal cell subset-related pathophysiologies. Furthermore, it has proven feasible to target transgene expression in neuronal subset-restricted fashions including also at least some of the cell populations that are selectively vulnerable in PD and DLB. Targeting transgene expression in catecholaminergic neurons including the dopaminergic neurons of the SN has been achieved[29,30]. Therefore, it should be possible to model pathophysiological effects of mutant and/or wild-type αSN (over)expression in this subset of cells, and studies in such mice are currently underway. We also have generated many different transgenic lines using mouse Thy1 regulatory sequences.

3. ANIMAL MODELS OF α-SYNUCLEINOPATHY.

Recently, we and others have demonstrated the development of α-synucleinopathy with Lewy-like features in transgenic mice[26-28] and transgenic Drosophila[25] following expression of either mutant or wild-type human αSN.

In our transgenic mice, as described here, αSN-encoding cDNAs were under control of mouse Thy1 regulatory sequences[31-33] that produced reliably neuron-specific expression and high protein levels in many central neurons of the mouse nervous system.

4. NEUROPATHOLOGY IN MICE EXPRESSING THE HUMAN α-SYNUCLEIN A53T MUTANT.

We produced six transgenic C57BL/6 lines, each expressing a Thy1αSNA53T transgene to a different level. Two lines (9813 and 9956) expressed transgene mRNA levels, that each resulted in a significant increase of total αSN protein levels in brain. In both of these lines, transgene expression occurred in neurons throughout the telencephalon, brainstem, and spinal cord. This was shown by *in situ* hybridization using cRNA probes that correspond either the full-length coding sequence of human αSN or to 111 bp encoding the carboxy-terminal 37 amino acids of human αSN. This latter probe is 87.5% homologous to mouse cDNA but has nucleotide differences occurring every 10-20 bp. As a result, this probe detects no signal above background in the non-transgenic C57BL/6 mouse brain and it can be used to monitor specifically the signal of the transgene.

In all the mice of lines 9813 and 9956, we observed an early-onset (> 3 weeks of age) and a progressive decline of motor performance. A more severe motor deficit was also seen in a single transgenic F1 male of a third independent founder (line 9832). This male died at an age of five weeks and the line was lost. Brain transgene mRNA levels in the line 9832 single F1 male superceded those detected in line 9813 (and all the other lines). The progressive decline in motor performance was quantified in a rotating rod experiment comparing groups of transgenic and non-transgenic littermate mice of line 9813, starting at age 40 days and up to age 200 days. The results showed a progressive decline in motor performance and a dramatic reduction in endurance of the transgenic mice to remain on the rotating rod.

This pronounced motor phenotype and the documented expression of Thy1-based transgenes in motor neurons[34] led us to examine more closely the pathological changes in these cells. In the anterior horns of the spinal cord approximately 80% of the motor neurons expressed the human αSN A53T mutant protein as shown by immunoreactivity for the human αSN-specific antibody LB509[9,35]. Specific to the transgenic mice, many of these cells showed diffuse perikaryal αSN staining, and some also Lewy-like pathology with pronounced ubiquitin immunoreactivity. Furthermore, staining with the Campbell-Switzer pyridine silver technique[36], a routine and sensitive method used to detect Lewy-type changes in human brain tissue[37,38], revealed intense staining. These results indicate that (rodent) motor neurons are susceptible to Lewy-like changes when expressing the A53T mutant of human αSN. Other changes associated with the development of the motor neuron pathology (and seen also in other affected brain regions) included astrocytic gliosis and microglial activation.

We also examined spinal roots and nerve fiber bundles in muscles that both immunostained for αSN. Axonal degeneration was apparent in spinal roots with nerve fibers showing breakdown and segmentation into ellipsoids of the myelin sheath. Skeletal muscles contained small angular fibers indicating neurogenic muscular atrophy, consistent with the denervation and neuropathology observed to varying degrees at neuromuscular junctions (NMJs) of gastrocnemius muscle taken from several line 9813 mice aged 6.5 months. Dependent on the individual mouse, 10-40% of the synapses were denervated and thinning of preterminal nerves and/or swellings were detected in at least 50% of innervated synapses. Pathological changes were also observed in two other muscles, the extensor digitorum longus, containing fast-twitch muscle fibers, and the soleus, which contains mostly slow-twitch muscle fibers. While the structure of the postsynaptic apparatus, visualized by α-bungarotoxin, was similar in transgenic and wild-type mice, the presynaptic motor neurons showed clear signs of degeneration. Transgenic NMJs as compared to C57BL/6 often showed a reduction in the neurofilament staining (indicating that motor neurons were about to retract from synapses), and discontinuous staining in more proximal regions of the nerve. The hypothesis that motor neurons abandon synaptic sites was supported in staining for the synaptic vesicle protein synaptophysin. The finding that neurodegeneration including denervation of NMJs is observed in all the muscles examined suggests that a universal mechanism of synapse maintenance is affected by transgene expression of human α-synuclein. Together with the neuropathological changes detected in central areas, including some that are involved in motor function, these findings provide a likely explanation for the observed reduction in complex motor performance.

To investigate pathological changes in the brain more closely, animals of both lines aged 12 weeks and older were examined by applying immunohistochemical techniques routinely used to assess Lewy pathology in human brain[1-3,7-10,14]. In transgenic mouse brains, many neurons in the telencephalon and brainstem showed intense α-synuclein staining in cell bodies and dendrites. This pattern was seen when using an antibody that specifically detects human but not mouse α-synuclein, and when using an antibody that detects both mouse and human α-synuclein. The staining patterns were in sharp contrast to the axonal and presynaptic distribution of endogenous mouse α-synuclein in the nervous system of non-transgenic mice.

Increased perikaryal and neuritic staining of neurons by αSN antibodies is one characteristic feature in the diseased human brain[4-10]. As compared to affected neurons in human PD brain, αSN staining in transgenic mouse brains showed heterogeneous changes in neurites, very similar to those observed in human brain[1-10]. Frequently observed changes included sausage-like enlargements of proximal and distal neuritic segments, thick or fine thread-like inclusions as well as beaded or spindle-shaped neurites. They were most prominent and frequent in areas including the nucleus centralis oralis pontis, the nucleus vestibularis lateralis, the deep cerebellar nuclei, the deep aspects of the tectal plate, and the motor nuclei in the spinal cord.

Immunostaining for ubiquitin (Ub) is also frequently used to visualize Lewy pathology in human brain[1-10,14,39]. In the transgenic mice, dystrophic neurites and cell bodies occasionally stained intensely for Ub. Curiously, ubiquitination was pronounced in some but not all of the mice with Lewy-like αSN immunopathology and the phenomenon was observed in both lines (9813 and 9956) and sexes. The stochastic fashion in which ubiquitination appeared was not caused by differences in transgene mRNA expression or strain genetic background (all mice are C57BL/6). Therefore, factors other than these play a role in this phenomenon. Ub-immunostained neurites displayed morphological features similar to those seen in human brains with Lewy pathology[1-10,14]. However, in transgenic mouse brains both Ub-immunostained neurons and neurites were less frequent when compared to αSN-positive cells with Lewy-like features. And like the latter, Ub-positive cells were most prominent and frequent in areas including the nucleus centralis oralis pontis, the nucleus vestibularis lateralis, the deep cerebellar nuclei, the deep aspects of the tectal plate, and the motor nuclei in the spinal cord. Also in human brains with Lewy pathology, Ub-immunopositive cells are less frequent as compared to αSN-immunopositive cells with Lewy pathology[1-6,39], suggesting that ubiquitination might be a late event in the development of the pathology in mouse and human. A quantification of the number of perikarya and neurites showing Ub versus αSN immunopathology was not attempted because unlike in human brains with Lewy pathology, the pattern and intensity of Ub-immunolabeling was very heterogeneous. In DLB, percentages of αSN-positive but Ub-negative structures ranged only from 2-10 % dependent on the brain region[39]. Morphologically, and as compared to classical Lewy bodies, the Ub-negative αSN-positive structures corresponded to less aggregated and less compact αSN inclusions like pale bodies[40]. αSN as compared to Ub antibodies thus showed higher sensitivity for identifying less aggregated and less compacted inclusions but little differences with respect to identifying classical Lewy bodies. In the mice we have not seen classical Lewy body-like structures. Perikarya and neurites contained less compact and more diffuse αSN immunopositive material. A significant proportion of αSN-immunopositive perikarya and neurites failed to show or had different degrees of punctate Ub staining. Only occasional cells and neurites showed a very intense Ub staining. Therefore, as compared to human brains with Lewy pathology, the majority of αSN positive inclusions in the mice may represent less compacted αSN conglomerates associated with pre- or early stages of ubiquitination. In human brains with Lewy pathology such as DLB, ubiquitinated structures represent mainly classical Lewy bodies and Lewy neurites whereas punctate Ub-pathology is much less frequently detected[39]. It should also be noted that it is not clear whether significant dynamic changes occur of ubiquitination in cells and neurites over time. Perhaps, cells and neurites showing significant Ub immunoreactivity at one given time point would show little or no reactivity at a later time point. Lastly, regulation of protein function by conjugation to Ub[41] is only one of several known Ub-like modification systems including e.g. SUMO-1/sentrin-1 and NEDD8[42,43]. Whether such Ub-like protein modifications also accumulate in αSN transgenic mice is currently under investigation.

Ultrastructural features of human αSN-positive neuronal inclusions in Thy1αSNA53T mice have been examined using conventional and immunoelectron microscopy. Neurons of transgenic but not C57BL/6 mice showed mainly αSN-containing fine granular deposits, unlike the predominantly fibrillar appearance of αSN immunoreactive inclusions seen in human brains with Lewy pathology[3,7,8].

5. NEUROPATHOLOGY IN TRANSGENIC MICE EXPRESSING WILD-TYPE HUMAN α-SYNUCLEIN.

The vast majority of human cases with Lewy pathology are idiopathic and αSN mutations in the coding region, other than A53T and A30P, have not been found in cohorts of familial or sporadic PD[44-47]. To test whether the pathology seen in the mice is unique for the A53T mutant of αSN, in which case it could bear little on idiopathic forms of the diseases, we also generated transgenic lines expressing wild-type human αSN under control of the mouse Thy1 regulatory sequences as for the A53T mutant. Mice of one of these lines (S969) showed a motor phenotype, and expressed brain transgene mRNA and protein levels similar to those seen in Thy1αSNA53T mice of line 9813. Brain and spinal cord contained many neurons with pronounced αSN immunostaining in cell perikarya and dendrites. Cells and neurites with Lewy-like features were prominent and frequent mainly in those areas of the central nervous system as reported for the Thy1αSNA53T mice. Gastrocnemius neuromuscular junctions showed in some regions over 50% denervation, and sprouting was often seen at the remaining endplates like in the A53T mice. In conclusion, like its A53T mutant, expression of wild-type human αSN can cause pathogenicity in central neurons. Masliah et al.[26] recently described several mouse lines of wild-type human αSN overexpression using the platelet-derived growth factor-β promoter (PDGFβ). Mice in all of these lines developed neuronal cytoplasmic and nuclear inclusions containing human αSN and Ub. Nuclear inclusions are atypical in sporadic PD and DLB[39] and perhaps these inclusions reflect a cell's attempt and/or failure to clear the αSN protein molecules via degradation pathways such as the proteasome pathway[48]. Interestingly, similar to what we have seen in Thy1αSN mice, PDGFβ-αSN mice were found to contain fine granular but no clear cut fibrillar αSN immunoreactive intracytoplasmic inclusions when using immunoelectron microscopy. 12-month-old mice of the high PDGFβ-αSN expressing line showed degeneration of dopaminergic terminals, decrements in the activity of tyrosine hydroxylase (TH), and an impairment in rotarod performance. It remains to be seen, whether these motor deficits are related to the small decrements in TH activity, as motor symptoms in PD patients and MPTP treated animals become apparent after 60-80% of dopaminergic neurons in the SN have been lost[49]. Furthermore, as we have shown in Thy1αSN transgenics, severe motor deficits can evolve in αSN transgenics without a significant expression of human αSN in the dopaminergic neurons.

In a recently described fruit fly model with α-synucleinopathy[25], overexpression of mutant and wild-type human αSN resulted in an adult-onset and progressive loss of dopaminergic neurons. These flies also developed a progressive and age-dependent locomotor dysfunction. Interestingly, the flies contained both granular and filamentous αSN immunoreactive structures with an overall apprearance that resembles Lewy structures seen in human PD brain. Also striking and despite a panneuronal expression of the transgenes, the flies showed a selective loss of dorsomedial but not other dopaminergic neurons. Expression of αSN specifically in non-dopaminergic cells of the developing and adult fly eye resulted in an age-dependent retinal degeneration. Thus also in the fly, like in our Thy1-based transgenic mice, and in human PD and related syndromes such as DLB, degenerative changes can occur in nondopaminergic cells.

6. GENERAL DISCUSSION

The few so far described transgenic mouse and fly models cover already a remarkable spectrum of the pathological changes that are seen in *post mortem* brain tissue of DLB, PD and LBVAD patients. These studies have also identified the A53T mutant of human αSN as well as its wild-type counterpart as pathogenic agents that can drive the formation of extranigral neuropathology in rodent or fly central neurons. The A53T mutation may therefore enhance but is not required for pathogenicity of the human protein. Expression of the human αSN A30P mutant has also revealed neuronal changes with Lewy-like features in both the mouse[28] and the fly[25]. Therefore, all three human αSN proteins can be pathogenic in neurons *in vivo*.

In each case, the expression of human αSN resulted in aberrant perikaryal and dendritic accumulations of human αSN in neurons, suggesting that this could play a central role in the development of Lewy-like pathology. In Thy1αSN mice (A53T and wt), human αSN expression also induced motor neuron degeneration with profound effects on axonal integrity, and denervation of NMJs in several muscles examined. Why such a phenotype was not seen in the so far generated Thy1αSNA30P mice remains to be investigated[28]. It is not yet clear whether this relates to properties of the specific mutant protein, to differences in expression patterns in motor neurons, and/or to differences in the αSN (non-coding) cDNA sequences that were used to generate the mice. In any case, our observations in the Thy1αSN mice suggest that αSN (over)expression affects a perhaps universal mechanism of synapse maintenance. The degeneration of dopaminergic terminals in PDGFβ-αSN mice also supports this notion[26].

Whether the three human αSN proteins affect one-and-the-same or different molecular mechanisms, and with similar or different thresholds for pathogenicity remains to be determined. Wild-type and mutant (A30P and A53T) αSN self aggregate and assemble into fibrils that resemble the ultrastructural elements of Lewy bodies[21]. Recent observations that the A30P mutant fibrillizes more slowly than its wild-type counterpart suggest that fibrils may not be the toxic moiety[21]. But both mutants oligomerize more rapidly that wild-type αSN potentially implicating oligomers and protofibrils in the pathogenesis. If so, the presence of fibrils and Lewy bodies could merely reflect a cell's attempt to detoxify damaged or mutant αSN.

Molecular mechanisms underlying neuropathologies with α-synucleinopathy likely involve altered interactions of mutated or otherwise modified αSN molecules with one or several molecules, other proteins and/or cellular organelles such as vesicles and membranes[50,51]. *In vitro* studies analysing the interaction of αSN and Aβ protein under physiological conditions suggested that αSN can bind to Aβ through specific sequences in the NAC domain as well as at the N-terminus[52] and thus might promote the aggregation of Aβ in Alzheimer patients[53]. A role of αSN in regulating phospholipase D2 has been suggested by Jenco and coworkers[54]. αSN shares more than 40% homology with 14-3-3 proteins, a family of molecular chaperones[55], thus it might have a similar function. In fact, αSN has been shown to dimerize with 14-3-3 proteins, and to co-immunoprecipitate with some proteins that interact with 14-3-3 proteins such as protein kinase C, BAD and extracellular regulated kinase. Another interaction that has been discovered involves the C-terminus of αSN with the microtubule binding domain of tau. This interaction seems to stimulate tau phosphorylation by protein kinase A, and therefore it could have influence on the stability of axonal microtubules[56]. Interestingly, microtubule-associated protein 1B (MAP1B) was recently shown to occur in cortical Lewy bodies and to bind to αSN filaments[57]. Direct binding studies revealed a high affinity interaction ($IC_{50} \sim$ 20nM) between αSN and MAP1B and this interaction appears to involve a region within the last C-terminal 45 amino acids of αSN. Another novel interacting protein, synphilin-1, has recently been discovered as a cytosolic partner for αSN using a yeast-two-hybrid approach. Synphilin-1 contains several protein-protein interaction domains such as ankyrin-like repeats and a coiled-coil domain. Synphilin-1 might act as an adaptor molecule anchoring αSN to intracellular proteins involved in vesicular transport and cytoskeletal functions. It also carries a ATP/GTP-binding motif. The finding that co-transfection of synphilin-1 with the NACP-fragment of αSN gave rise to cytoplasmic eosinophilic inclusions[58], and that synphilin-1 is a component of Lewy bodies in PD patients[59] could suggest that an interaction of synphilin-1 with αSN represents more than just an epiphenomenon in α-synucleinopathies. Finally, αSN is predominantly a presynaptic protein and thus requires axonal transport to reach its destination. Axonal transport of αSN occurs via all axonal transport rate components, the fast component and the slow components a and b, with component b predominating[60]. Given the abundance of αSN in brain, improperly folded, modified or mutated αSN could likely interfere with each one of the different transport mechanisms and perturb the transport of different single and/or multi-molecular[61](e.g. preformed synaptic complexes) entities. Dependent on the mode of insult, perhaps certain axonal transport components are affected more. But eventually, αSN-mediated interference with any of these axonal transport components could disturb axonal and synaptic integrity. A role for αSN in presynaptic terminals was recently proposed. Findings in αSN knockout (-/-) mice indicate that the wild-type αSN protein might be involved in regulating dopamine (DA) neurotransmission[62]. αSN-/- mice did not show marked morphological or pathological brain abnormalities but suffered from significant neurochemical, electrophysiological and behavioral deficits. The mice exhibited a reduction in total striatal DA (18%), faster recovery from paired stimulus depression (35% more rapid than wt littermates) and they showed a significantly reduced locomotor response to amphetamine-stimulation. It remains to be seen whether mutant or otherwise pathologically modified αSN can interfere with these or other wild-type αSN functions. Although its localisation in presynaptic terminals[63], its association with vesicles[50,64,65], and the developmental expression of the αSN gene has led to suggestions of roles of the protein in synaptogenesis[63], one recent study on the characterization of αSN expression in primary hippocampal neuronal cultures revealed delayed expression of αSN after synapse formation[66]. Together with a lack of gross morphological abnormalities related to synaptogenesis in αSN-/- mice, most findings therefore do not seem to suggest a critical role of αSN in this process. But downregulation of αSN expression by antisense-oligonucleotides did cause a significant and selective reduction in the size of the distal presynaptic vesicular pool. This could suggest a role of αSN in modulating synaptic function in the normal brain by influencing specific synaptic vesicle pools[66]. However, also of note, expression of αSN occurs in several non-neuronal cell types (e.g. hematopoietic cells) suggesting functions of αSN outside the nervous system and other than in neurotransmission[67]. It will be interesting to see if and what kind of αSN-related changes occur in peripheral tissues of mice and patients with central α-synucleinopathies.

Although not many reports have described secondary modifications of αSN, the protein has been found to be constitutively phosphorylated in stably transfected 293 and PC12 cells[68]. In these studies, Ser-129 was the major phosphoacceptor site and casein kinase I and II were likely responsible phosphotransferases whereas Ca2+/calmodulin-dependent protein kinase II, which strongly phosphorylates βSN[69], only moderately phosphorylated αSN[68]. The physiological role of αSN phosphorylation/dephosphorylation with respect to the protein's functions, molecular interactions, and/or aggregation remains an interesting area of future research, in particular, if correlates could be found with its neuropathological properties. And since the αSN-encoding gene is infrequently mutated in its coding sequence, it will be key to identify and study cellular processes that facilitate αSN protein accumulation. Apart from (over)expression of wild-type (or mutant) human αSN in transgenic mice and flies, it has also been shown to occur following a specific insult[70,71]. Protein degradation pathways such as the proteasome-ubiquitin pathway may also play a key role in determining the protein's accumulation in pathological situations[72,73].

An interesting feature of the pathology in the transgenic mice, the transgenic flies, and in human diseases with Lewy pathology[1-3,74,75], is a selective and topographically-confined vulnerability of neurons to develop Lewy-like pathology. In the Thy1αSN mice, prominently affected areas included the nucleus centralis oralis pontis, the nucleus vestibularis lateralis, the deep cerebellar nuclei, the deep aspects of the tectal plate, and motor nuclei in the spinal cord. Of course, this pattern is dictated *a priori* by the properties of the Thy1 expression cassette. This cassette yielded widespread neuronal expression of human αSN. But pathology was most pronounced in certain brain regions suggesting that some neurons are more vulnerable. Perhaps modulators of α-synucleinopathy vary regionally in the brain. The Thy-1 cassette generally fails to express in dopaminergic neurons of the *substantia nigra pars compacta* (SNc) (S. Bischoff and HvdP, unpublished observations) and we confirmed this to be the case also in the A53T transgenic mice (data not shown). In the human brain, the dopaminergic cells of the SNc appear most sensitive to develop Lewy pathology and the resulting nigral lesions are primarily involved in the clinical symptoms of PD[1-10,76-79]. Furthermore, expression in dopaminergic neurons of PDGFβ-αSN mice showed that subsets of these but likely not all dopaminergic cells are highly prone to develop Lewy-like pathology[26]. Similar observations were reported for the transgenic flies[25], suggesting that sparing and vulnerability of subsets of dopaminergic neurons is perhaps a common and a very intriguing feature in the various animal models and in patients with sporadic PD.

But in addition, extranigral Lewy pathology is very common in both PD and DLB brains, and it is suspected to contribute to the variable cognitive and neuropsychiatric features in these diseases[1,14,74,75,80-87]. By excluding transgene effects in dopaminergic neurons of the SNc, Thy1αSN mice seem particularly useful to address and model extranigral aspects of the pathology as it is seen in PD and DLB whereas a new generation of mice with specific expression of αSN proteins in SNc may be useful to specifically address dopaminergic (and other catecholaminergic) pathophysiological changes.

The remarkable spectrum of Lewy-like pathological changes seen in the transgenic mouse models reported thus far included subsets of central neurons showing αSN-stained perikarya, dendrites, and Lewy-like neurites, and a smaller subset of these cells staining for ubiquitin. In the transgenic flies, ubiquitin staining was not reported, perhaps because of technical difficulties. In our Thy1αSN mice, we observed no pathological staining when using antibodies that detect paired helical filament and hyperphosphorylated tau, neurofilaments, or when using stains to detect Aβ plaque-like pathology[32].

Like in human PD and DLB brains[1-6], ubiquitination in transgenic mouse neurons appears to be a late modification in cells with αSN pathology. Ubiquitination was evident in some but not in all aged-matched female and male Thy1αSN mice[27] and this phenomenon was observed in two Thy1A53T lines (and a Thy1αSNwt line). When it was seen, it was often pronounced in regionally confined subsets of neurons within the same animal and it was detected mainly in those brain regions where we had seen the most pronounced cellular and neuritic α-synucleinopathy. Since all the mice were C57BL/6, and no variation in transgene expression was detected between individuals or sexes within a line, factors other than these seem to account for ubiquitination occurring in a stochastic fashion. At this point it is not clear whether mice that show significant ubiquitination have benefit over those that do not, or whether it is associated with even more detrimental effects in the cells where it occurs. A major role of protein ubiquitination is thought to be directing target proteins for degradation by the proteasome. The detection of excessive ubiquitination could suggest the existence of αSN protein that is resistant to proteolysis and perhaps it reflects an accumulation of blocked intermediates in the turnover pathway that is used by the wild-type protein. However, it could also reflect attempts of cells to degrade protein whose turnover is compromised by mutation and/or secondary modification. In diseases characterized by proteins suffering from polyglutamine expansions, there is accumulating evidence in favour of the latter hypothesis[88].

Using transmission and immunoelectron microscopy, human αSNA53T immunoreactive granular material was detected in cytoplasm, dendrites, and occasionally in axons of Thy1A53T mice. Granular material with αSN has been seen in human tissue with Lewy pathology and in partially purified human Lewy bodies[9,89,90]. We did not detect the typical αSN decorated Lewy body filaments as seen in material extracted from human DLB cingulate cortex[5], in tissue sections from human PD and DLB brain[9,89,90], and in synthetic αSN filaments[20,22-24,91,92]. This could relate to insufficient ageing of the mice and/or species differences. Others recently reported granular but no fibrillar material in mice expressing wild-type human αSN[26]. Interestingly, in transgenic Drosophila fibrillar material was detected[25]. These findings are intriguing because the identity of the pathogenic species in diseases with Lewy pathology and its relationship to the αSN containing fibril has not been elucidated, and it is not known whether Lewy bodies in human diseases are cytotoxic, harmless side products, or markers of cell damage. Irrespective, and likely more important, are the several other striking similarities seen between the pathology in human and the transgenic mice.

All three (wt, A30P and A53T) αSN proteins produce nonfibrillar oligomers[21] that perhaps are critical in pathogenesis, whereas fibrils and/or Lewy bodies could represent harmless epiphenomena.

Aging and/or species differences may also account for the prominent, diffuse, ubiquitin staining seen in the cytoplasm of mouse but not human neuronal cell bodies with Lewy-like pathology. Ubiquitin-staining of dystrophic neurites is remarkably similar in both species, but in human cell perikarya, the staining is usually confined to Lewy bodies[1-6,14].

The role of immune and inflammatory mechanisms in the pathogenesis of diseases with Lewy pathology, and PD in particular, is still a matter of debate. Immunological abnormalities have been reported in various brain areas and in peripheral immune parameters[49,93]. In the substantia nigra of PD brains, a huge increase in the density of IFN-γ, IL-β, and TNF-α-expressing glial cells has been reported[94]. The same authors have also pointed out a putative role for a CD23-(low affinity IgE receptor)-dependent activation pathway of inducible nitric oxide synthase (iNOS) and proinflammatory mediators. In 1-methyl-4-phenyl-1,2,3,6-tetrahydropyridine (MPTP)-induced parkinsonism there is emerging evidence for neuroinflammation in the pathogenesis[95]. Inflammatory mechanisms are often accompanied by increased oxidative stress. Several changes detected in PD brains indicate an increase in oxidative stress such as glutathione depletion, iron deposition, increased markers of lipid peroxidation, oxidative DNA damage, and protein oxidation[49]. Furthermore, decreased expression and activity of the mitochondrial complex I in the midbrain has been reported to occur specifically in PD and not in other neurodegenerative diseases (see Dunnett and Bjorklund[49], and references therein). In the Thy1αSN mice, changes associated with the development of the motor neuron pathology in spinal cord and the pathology seen in other affected brain regions included astrocytic gliosis and microglial activation, suggesting that neuroinflammation could play a role in this model[27]. Irrespective of inflammatory processes being primary or secondary to neurodegeneration, inflammatory components might be attractive targets for therapeutic intervention since they likely contribute to pathogenesis. Anti-inflammatory agents have been shown to protect dopaminergic neurons against degeneration in the MPTP animal model of PD[96,97].

MPTP appears to elicit dopaminergic neuron degeneration through inhibition of complex I and the formation of superoxide anion. The brain consumes large amounts of oxygen and is therefore particularly vulnerable to oxidative damage and dependent on antioxidant defenses. Deficiencies in these mechanisms, such as a lack of the cellular enzyme glutathione peroxidase (GSHPx) in GSHP-/- mice resulted in increased vulnerability to MPTP[98]. GSHPx is present in both neurons and astrocytes, cytosol and mitochondria, and the enzyme is a major player in antioxidant defense by converting H_2O_2 to H_2O and by detoxifying peroxynitrite, ONOO-[99]. Inflammatory processes often result in increased nitric oxide synthase activity (NOS) and nitric oxide (NO) production (see Hunot et al.[94], and references therein), resulting in increased peroxynitrite formation. In turn, this may result in oxidative injury to dopaminergic neurons, perhaps involving poly (ADP-ribose) polymerase as a major contributor[100]. Both neuronal NOS[101-104] and inducible (i) NOS[95,105] appear to play a key role in stimulating dopaminergic neurodegeneration in the MPTP model. In vitro and in vivo data indicate that oxidative stress can induce α-SN aggregation[70]. A putative role for nitrating agents in diseases with α-synucleinopathy is strenghtened by recent findings showing that αSN inclusions in PD, DLB and MSA brains are strongly labeled with antibodies specific to 3-nitrotyrosine[106] and antibodies specifically detecting nitrated αSN[107]. Interestingly, exposure of human recombinant αSN (containing four tyrosines) but for example not βSN (which has similarly placed tyrosines in C-terminus) to nitrating agents induced formation of tyrosine nitrated αSN oligomers that were highly stabilized due to covalent cross-linking via oxidation of a most likely C-terminal tyrosine to form o,o'-dityrosine[107]. Noteworthy, βSN and γSN pathology was recently reported in diseases with Lewy pathology, but it is clearly much more confined regionally as compared to α-synucleinopathy in the diseased brain and predominantly hippocampal. Also, immunohistologically it is very distinct from αSN-containing inclusions that lack βSN and γSN[108]. Finally, enhanced vulnerability to oxidative stress following exposure to H_2O_2 has been observed in SH-Sy5Y cells transfected with C-terminally truncated αSN proteins, particularly the 1-120 residue αSN protein, and the two mutants (A53T and A30P) linked to familial forms of PD[109].

Despite intensive research efforts during recent years, fundamental questions remain regarding the etiology and pathogenesis of diseases with Lewy pathology. Only in rare cases of PD has neuronal injury been linked to genetic factors, whereas idiotypic forms have been linked to infectious agents, exogenous toxins, genetic factors, excitatory amino acids, endogenous reactive oxygen species, or other metabolic disturbances such as a lack of trophic support. Interestingly, even the enzyme TH remains a suspect in PD mechanisms related to oxidative stress. The TH locus is not a major genetic risk factor in PD but L-DOPA is effectively oxidized by mammalian TH in vitro, and therefore TH could contribute to cytotoxic effects of L-DOPA[110]. The recent and novel pathophysiological findings in a first series of transgenic mouse and fly models now provide compelling evidence for a role of αSN in neuropathologies with α-synucleinopathy. There is still a lot more to learn with respect to mechanisms involved, and also there is certainly room to improve and refine the animal models. It is hoped that they will provide useful means to address fundamental aspects of disorders with α-synucleinopathy, and to test and discover novel routes for therapy.

REFERENCES

1. Forno LS (1996) Neuropathology of Parkinson's disease. J Neuropathol Exp Neurol 55: 259-271.
2. Ince PG, Perry EK, Morris CM (1998) Dementia with Lewy bodies. A distinct non-Alzheimer dementia syndrome. Brain Pathol 8: 299-324.
3. Takeda A, Mallory M, Sundsmo M et al. (1998) Abnormal accumulation of NACP/α-synuclein in neurodegenerative disorders. Am J Pathol 152: 367-372.
4. Irizarry MC, Growdon W, Gomez-Isla T et al. (1998) Nigral and cortical Lewy bodies and dystrophic nigral neurites in Parkinson's disease and cortical Lewy body disease contain α-synuclein immunoreactivity. J Neuropathol Exp Neurol 57: 334-337.
5. Spillantini MG, Crowther RA, Jakes R et al. (1998) α-Synuclein in filamentous inclusions of Lewy bodies from Parkinson's disease and dementia with Lewy bodies. Proc Natl Acad Sci USA 95: 6469-6473.
6. Braak H, Sandmann-Keil D, Gai W et al. (1999) Extensive axonal Lewy neurites in Parkinson's disease: a novel pathological feature revealed by α-synuclein immunocytochemistry. Neurosci Lett 265: 67-69.
7. Wakabayashi K, Matsumoto K, Takayama K et al. (1997) NACP, a presynaptic protein, immunoreactivity in Lewy bodies in Parkinson's disease. Neurosci Lett 239: 45-48.
8. Spillantini MG, Schmidt M, Lee VM et al. (1997) α-Synuclein in Lewy bodies. Nature 388: 839-840.
9. Baba M, Nakajo S, Tu PH et al. (1998) Aggregation of α-synuclein in Lewy bodies of sporadic Parkinson's disease and Dementia with Lewy bodies. Am J Pathol 152: 879-884.
10. Mezey E, Dehejia A, Harta G et al. (1998) α-Synuclein is present in Lewy bodies in sporadic Parkinson's disease. Mol Psychiatry 3: 493-499.
11. Den Hartog WA, Bethlem J (1960) The distribution of Lewy bodies in the central and autonomic nervous systems in idiopathic paralysis agitans. J Neurol Neurosurg Psychiat 23: 283-290.
12. Arai T, Ueda K, Ikeda K et al. (1999) Argyrophilic glial inclusions in the midbrain of patients with Parkinson's disease and diffuse Lewy body disease are immunopositive for NACP/alpha-synuclein. Neurosci Lett 259: 83-86.
13. Giasson BI, Jakes R, Goedert M et al. (2000) A panel of epitope-specific antibodies detects protein domains distributed throughout human α-synuclein in Lewy bodies of Parkinson's disease. J Neurosci Res 59: 528-533.
14. Gai WP, Blessing WW, Blumbergs PC (1995) Ubiquitin-positive degenerating neurites in the brainstem in Parkinson's disease. Brain 118: 1447-1459.
15. Lowe J, McDermott H, Landon M et al. (1990) Ubiquitin carboxyl-terminal hydrolase (PGP9.5) is selectively present in ubiquinated inclusion bodies characteristic of human degenerative disorders. J Pathol 161: 153-160.

16. Li K., Ito H, Tanaka K et al. (1997) Immunocytochemical co-localization of the proteasome in ubiquitinated structures in neurodegenerative diseases and the elderly. J Neuropathol Exp Neurol 56: 125-131.
17. Pollanen MS, Dickson DW, Bergeron C (1993) Pathology and biology of the Lewy body. J Neuropath Exp Neurol 52: 183-191.
18. Polymeropoulos MH, Lavedan C, Leroy E et al. (1997) Mutation in the α-synuclein gene identified in families with Parkinson's disease. Science 276: 2045-2047.
19. Krüger R, Kuhn W, Müller T et al. (1998) Ala30Pro mutation in the gene encoding α-synuclein in Parkinson's disease. Nat Gen 18: 106-108.
20. Conway KA, Harper JD, Lansbury PT (1998) Accelerated in vitro fibril formation by a mutant α-synuclein linked to early-onset Parkinson's disease. Nat Med 4: 1318-1320.
21. Conway KA, Lee S-J, Rochet J-C et al. (2000) Acceleration of oligomerization, not fibrillization, is a shared property of both α-synuclein mutations linked to early-onset Parkinson's disease: Implications for pathogenesis and therapy. Proc Natl Acad Sci USA 97: 571-576.
22. Giasson BI, Uryu K, Trojanowski JQ et al.(1999) Mutant and wild type human alpha-synucleins assemble into elongated filaments with distinct morphologies in vitro. J Biol Chem 274: 7619-7622.
23. Narhi L, Wood SJ, Steavenson S et al. (1999) Both familial Parkinson's disease mutations accelerate alpha-synuclein aggregation. J Biol Chem 274: 9843-9846
24. Wood SJ, Wypych J, Steavenson S et al. (1999) alpha-synuclein fibrillogenesis is nucleation-dependent. Implications for the pathogenesis of Parkinson's disease. J Biol Chem 274: 19509-19512.
25. Feany MB, Bender WW (2000) A Drosophila model of Parkinson's disease. Nature 404: 394-398.
26. Masliah E, Rockenstein E, Veinbergs I et al. (2000) Dopaminergic loss and inclusion body formation in α-synuclein mice: implications for neurodegenerative disorders. Science 287: 1265-1269.
27. van der Putten H, Wiederhold K-H, Probst A et al. (2000). Neuropathology in mice expressing human α-synuclein. J Neurosci 20: 6021-6029.
28. Kahle PJ, Neumann M, Ozmen L et al. (2000) Physiology and pathophysiology of α-synuclein: cell culture and transgenic animal models based on a Parkinson's disease-associated protein. In: Molecular Genetics of Dementia, Proceedings of the ninth meeting of the international study group on the pharmacology of memory disorders associated with aging, Zuerich (Growdon JH, Wurtman RJ, Corkin SS, Nitsch RM eds), pp45-54: Cambridge MA, USA.
29. Min N, Joh TH, Kim KS et al. (1994) 5'upstream DNA sequence of the rat tyrosine hydroxylase gene directs high-level and tissue-specific expression to catecholaminergic neurons in the central nervous system of transgenic mice. Mol Brain Res 27: 281-289.
30. Liu J, Merlie JP, Todd RD et al. (1997) Identification of cell type-specific promoter elements associated with the rat tyrosine hydroxylase gene using transgenic founder analysis. Mol Brain Res 50: 33-42.
31. Luethi A, van der Putten H, Botteri FM et al. (1997) An endogenous serine protease inhibitor alters epileptic activity and hippocampal long-term potentiation. J Neurosci 17: 4688-4699.
32. Sturchler-Pierrat C, Abramowski D, Duke M et al. (1997) Two amyloid precursor protein transgenic mouse models with Alzheimer disease-like pathology. Proc Natl Acad Sci USA 94: 13287-13292.

33. Wiessner C, Allegrini PR, Rupalla K et al. (1999) Neuron-specific transgene expression in brain of mouse bcl-X_L but not human bcl-2 gene reduces lesion size following permanent MCAO. Neurosci Lett 268: 119-122.
34. Aigner L, Arber S, Kapfhammer JP et al. (1995) Overexpression of the neuronal growth-associated protein GAP-43 induces nerve sprouting in the adult nervous system of transgenic mice. Cell 83: 269-278.
35. Jakes R, Crowther RA, Lee VMY et al. (1999) Epitope mapping of LB509, a monoclonal antibody directed against human α-synuclein. Neurosci Lett 269: 13-16.
36. Campbell SK, Switzer RL, Martin TL (1987) Alzheimer's plaques and tangles: A controlled and enhanced silver staining method. Soc Neurosci Abstr 13: 678.
37. Braak E, Braak H (1999) Silver staining method for demonstrating Lewy bodies in Parkinson's disease and argyrophilic oligodendrocytes in multiple system atrophy. J Neurosci Methods 87: 111-115.
38. Sandmann-Keil D, Braak H, Okochi M et al. (1999) Alpha-synuclein immunoreactive Lewy bodies and Lewy neurites in Parkinson's disease are detectable by an advanced silver-staining technique. Acta Neuropathol 98: 461-464.
39. Gómez-Tortosa E, Newll K, Irizarry MC et al. (2000) α-synuclein immunoreactivity in dementia with Lewy bodies: morphological staging and comparison with ubiquitin immunostaining. Acta Neuropathol 99: 352-357.
40. Dale G, Probst A, Luthert P et al. (1992) Relationships between Lewy bodies and pale bodies in Parkinson's disease. Acta Neuropathol 83: 525-529.
41. Hershko A, Chiechanover A (1998) The ubiquitin system. Ann Rev Biochem 67: 425-479.
42. Tanaka K, Suzuki T, Chiba T (1998) The ligation systems for ubiquitin and ubiquitin-like proteins. Mol Cells 8: 503-12
43. Yeh ETH, Gong L, Kamitani T (2000) Ubiquitin-like proteins: new wines in new bottles. Gene 248: 1-14.
44. Chan P, Tanner CM, Jiang X et al. (1998) Failure to find the alpha-synuclein gene missense mutation (G209A) in 100 patients with younger onset Parkinson's disease. Neurol 50: 513-514.
45. Farrer M, Wavrant-De Vrieze F, Crook R et al. (1998) Low frequency of alpha-synuclein mutations in familial Parkinson's disease. Ann Neurol 43: 394-397.
46. Group T (1998) Alpha-synuclein gene and Parkinson's disease. Science 279: 1116-1117.
47. Vaughan J, Durr A, Tassin J et al. (1998) The alpha-synuclein Ala53Thr mutation is not a common cause of familial Parkinson's disease: a study of 230 European cases. Ann Neurol 44: 270-273.
48. Lee DH, Goldberg AL (1998) Proteasome inhibitors: valuable new tools for cell biologists. Trends Cell Biol 8: 397-403.
49. Dunnett SB, Björklund A (1999) Prospects for new restorative and neuroprotective treatements in Parkinson's disease. Nature 399: A32-A39.
50. Jensen PH, Nielsen MS, Jakes R et al. (1998) Binding of α-synuclein to brain vesicles is abolished by familial Parkinson's disease mutation. J Biol Chem 273: 26292-26294.
51. McLean PJ, Kawamata H, Ribich S et al. (2000) Membrane association and protein conformation of alpha-synuclein in intact neurons: Effect of Parkinson's disease-linked mutations. J Biol Chem 275: 8812-8816.
52. Hashimoto M, Masliah E (1999) Alpha-synuclein in Lewy body disease and Alzheimer's disease. Brain Pathol 9: 707-720.

53. Yoshimoto M, Iwai A, Kang D et al. (1995) NACP, the precursor protein of non-amyloid β/A4 protein (Aβ) component of Alzheimer disease amyloid, binds Aβ and stimulates Aβ aggregation. Proc Natl Acad Sci USA 92: 9141-9145.
54. Jenco JM, Rawlingson B, Daniels B et al. (1998) Regulation of phospholipase D2: selective inhibition of mammalian phospholipase D isoenzymes by α- and β-synucleins. Biochemistry 37: 4901-4909.
55. Ostrerova N, Petrucelli L, Farrer M et al. (1999) α-synuclein shares physical and functional homology with 14-3-3 proteins. J Neurosci 19: 5782-5791.
56. Jensen PH, Hager H, Nielsen MS et al. (1999) α-synuclein binds to tau and stimulates the protein kinase A-catalysed tau phosphorylation of serine residues 262 and 356. J Biol Chem 274: 25481-25489.
57. Jensen PH, Isalm K, Kenney J et al. (2000) Microtubule-associated protein 1B is a component of cortical Lewy bodies and binds alpha-synuclein filaments. J Biol Chem 275: 21500-21507.
58. Engelender S, Kaminsky Z, Guo X et al. (1999) Synphilin-1 associates with α-synuclein and promotes the formation of cytosolic inclusions. Nature 22: 110-114.
59. Wakabayashi K, Engelender S, Yoshimoto M et al. (2000) Synphilin-1 is present in Lewy bodies in Parkinson's disease. Ann Neurol 47: 521-523.
60. Jensen PH, Li J-Y, Dahlstrom A et al. (2000) Axonal transport of synucleins is mediated by all rate components. Eur J Neurosci 11: 3369-3376.
61. Roos J, Kelly RB (2000) Preassembly and transport of nerve terminals: a new concept of axonal transport. Nature 3: 415-417.
62. Abeliovich A, Schmitz Y, Farias I et al. (2000) Mice lacking α-synuclein display functional deficits in the nigrostriatal dopamine system. Neuron 25: 239-252.
63. Hsu LJ, Mallory M, Xia Y et al. (1998) Expression pattern of synucleins (non-Aβ component of Alzheimer's disease amyloid precursor protein/α-synuclein) during murine brain development. J Neurochem 71: 338-344.
64. Irizarry MC, Kim T-W, McNamara M et al.(1996) Characterization of the precursor protein of the non-Aβ component of senile plaques (NACP) in the human central nervous system. J Neuropathol Exp Neurol 55: 889-895.
65. Davidson WS, Jonas A, Clayton DF et al.(1998) Stabilization of alpha-synuclein secondary structure upon binding to synthetic membranes. J Biol Chem 273: 9443-9449.
66. Murphy DD, Rueter SM, Trojanowski JQ et al. (2000) Synucleins are developmentally expressed, and α-synuclein regulates the size of presynaptic vesicular pool in primary hippocampal neurons. J Neurosci 20: 3214-3220.
67. Shin EC, Cho SE, Lee DK et al. (2000) Expression patterns of alpha-synuclein in human hematopoietic cells and in Drosophila at different developmental stages. Mol Cells 10: 65-70.
68. Okochi M, Walter J, Koyama A et al. (2000) Constitutive phosphorylation of the Parkinson's disease associated α-synuclein. J Bio Chem 275: 390-397.
69. Nakajo S, Tsukada K, Omata K et al. (1993) A new brain-specific 14-kDa protein is a phosphoprotein. Its complete amino acid sequence and evidence for phosphorylation. Eur J Biochem 217: 1057-1063.
70. Kowall NW, Hantraye P, Brouillet E et al. (2000) MPTP induces alpha-synuclein aggregation in the substantia nigra of baboons. Neuroreport 11: 211-213.
71. Vila M, Vukosavic S, Jackson-Lewis V et al. (2000) Alpha-synuclein up-regulation in substantia nigra dopaminergic neurons following administration of the parkinsonian toxin MPTP. J Neurochem 74: 721-729.

72. Bennett MC, Bishop JF, Leng Y et al. (1999) Degradation of alpha-synuclein by proteasome. J Biol Chem 274: 33855-33858.
73. Leroy E, Boyer R, Auburger G et al. (1998) The ubiquitin pathway in Parkinson's disease. Nature 395: 451-452.
74. Braak H, Braak E, Yilmazer D et al. (1996) Pattern of brain destruction in Parkinson's and Alzheimer's diseases. J Neural Transm 103: 455-490.
75. Braak H, Rüb U, Sandmann-Keil D et al. (2000) Parkinson's disease: affection of brain stem nuclei controlling premotor and motor neurons of the somatomotor system. Acta Neuropathol 99: 489-495.
76. Lewy FH (1912) Paralysis agitans. I. Pathologische Anatomie. In: Lewandowski M (ed) Handbuch der Neurologie, vol III. Spinger, Berlin, pp 920-933.
77. Ohama E, Ikuta F (1976) Parkinson's disease: distribution of Lewy bodies and monoamine neuron system. Acta Neuropathol 34: 311-319.
78. Fearnley JM, Lees AJ (1991) Ageing and Parkinson's disease: substantia nigra regional selectivity. Brain 114: 2283-2301.
79. Gibb WRG, Lees AJ (1991) Anatomy, pigmentation, ventral and dorsal subpopulations of the substantia nigra, and differential cell death in Parkinson's disease. J Neurol Neurosurg Psychiatry 54: 388-396.
80. Greenfield JG, Bosanquet FD (1953) The brain-stem lesions in parkinsonism. J Neurol Neurosurg Psychiatry 16: 213-226.
81. Halliday GM, Li YW, Blumbergs PC et al. (1990) Neuropathology of immunohistochemically identified brainstem neurons in Parkinson's disease. Ann Neurol 27: 373-385.
82. Jellinger K (1991) Pathology of Parkinson's disease. Changes other than the nigrostriatal pathway. Mol Chem Neuropathol 14: 153-197.
83. Delwaide PJ, Pepin JL, Maertens de Noordhout A (1993) The audiospinal reaction in parkinsonian patients reflects functional changes in reticular nuclei. Ann Neurol 33: 63-69.
84. Braak H, Braak E, Yilmazer D et al. (1994) Amygdala pathology in Parkinson's disease. Acta Neuropathol 88: 493-500.
85. Braak H, Braak E, Yilmazer D et al. (1995) Nigral and extranigral pathology in Parkinson's disease. J Neural Transm 46: 15-31.
86. Nieuwenhuys R (1996) The greater lymbic system, the emotional motor system and the brain. Prog Brain Res 107: 551-580.
87. Gómez-Tortosa E, Irizarry MC, Gómez-Isla T et al. (2000) Clinical and neuropathological correlates of dementia with Lewy bodies. In: Molecular Genetics of Dementia, Proceedings of the ninth meeting of the international study group on the pharmacology of memory disorders associated with aging, Zuerich (Growdon JH, Wurtman RJ, Corkin SS, Nitsch RM eds), pp17-23: Cambridge MA, USA.
88. Ferrigno P, Silver PA (2000) Polyglutamine expansions: proteolysis, chaperones, and the dangers of promiscuity. Neuron 26: 9-12.
89. Arima K, Uéda K, Sunohara N et al. (1998) Immunoelectron microscopic demonstration of NACP/α-synuclein-epitopes on the filamentous components of Lewy bodies in Parkinson's disease and in dementia with Lewy bodies. Brain Res 808: 93-100.
90. Wakabayashi K, Hayashi S, Kakita A et al. (1998) Accumulation of α-synuclein/NACP is a cytopathological feature common to Lewy body disease and multiple system atrophy. Acta Neuropathol 96: 445-452.

91. Crowther RA, Jakes R, Spillantini MG et al. (1998) Synthetic filaments assembled from C-terminally truncated α-synuclein. FEBS Lett 436: 309-312.
92. El-Agnaf OM, Jakes R, Curran MD et al.(1998) Effects of the mutations Ala30 to Pro and Ala53 to Thr on the physical and morphological properties of alpha-synuclein protein implicated in Parkinson's disease. FEBS Lett 440: 67-70.
93. Kuhn W, Müller T, Nastos I et al. (1997) The neuroimminue hypothesis in Parkinson's disease. Rew Neurosci 8: 29-34.
94. Hunot S, Dugas N, Faucheux B et al. (1999) FcεRII/CD23 is expressed in Parkinson's disease and induces, in vitro, production of nitric oxide and tumor necrosis factor-α in glial cells. J Neurosci 19: 3440-3447.
95. Liberatore GT, Jackson-Lewis V, Vukosavic S et al. (1999) Inducible nitric oxide synthase stimulates dopaminergic neurodegeneration in the MPTP model of Parkinson disease. Nat Med 5: 1403-1409.
96. Matsuura K, Kabuto H, Makino H et al. (1996) Cyclosporin A attenuates degeneration of dopaminergic neurons induced by 6-hydroxydopamine in the mouse brain. Brain Res 733: 101-104.
97. Boireau A, Bordier F, Dubedat P et al. (1997) Thalidomide reduces MPTP-induced decrease in striatal dopamine levels in mice. Neurosci Lett 234: 123-126.
98. Klivenyi P, Andreassen OA, Ferrante RJ et al. (2000) Mice deficient in cellular glutathione peroxidase show increased vulnerability to malonate, 3-nitropropionic acid, and 1-methyl-4-phenyl-1,2,3,6-tetrahydropyridine. J Neurosci 20: 1-7.
99. Sies H, Sharov VS, Klotz LO et al. (1997) Glutathione peroxidase protects against peroxynitrite-mediated oxidations. J Biol Chem 272: 27812-27817.
100. Mandir AS, Przedborski S, Jackson-Lewis V et al. (1999) Poly(ADP-ribose) polymerase activation mediates 1-methyl-4-phenyl-1,2,3,6-tetrahydropyridine (MPTP)-induced parkinsonism. Proc Nat Acad Sci USA 96: 5776-5779.
101. Schultz JB, Matthews RT, Muquit MMK et al. (1995) Inhibition of neuronal nitric oxide synthase by 7-nitroindazole protects against MPTP-induced neurotoxicity in mice. J Neurochem 64: 936-939.
102. Przedborski S et al. (1996) Role of neuronal nitric oxide in MPTP (1-methyl-4-phenyl-1,2,3,6-tetrahydropyrirdine)-induced dopaminergic neurotoxicity. Proc Natl Acad Sci USA 93: 4565-4571.
103. Hantraye P, Brouillet E, Ferrante R et al. (1996) Inhibition of neuronal nitric oxide synthase prevent MPTP-induced parkinsonism in baboons. Nat Med 2: 1017-1021.
104. Matthews RT, Yang LC, Beal MF (1997) S-methylthiocitrulline, a neural nitric oxide synthase inhibitor, protects against malonatc and MPTP neurotoxicity. Exp Neurol 143: 282-286.
105. Grünewald T, Flint Beal M (1999) NOS knockouts and neuroprotection. Nat Med 5: 1354-1355.
106. Good PF, Hsu A, Werner P et al. (1998) Protein nitration in Parkinson's disease. J Neuropathol Exp Neurol 57: 338-342.
107. Souza JM, Giasson BI, Chen Q et al. (2000) Dityrosine cross-linking promotes formation of stable α-synuclein polymers: implication of nitrative and oxidative stress in pathogenesis of neurodegenerative synucleinopathies. J Biol Chem 275: 18344-18349.
108. Galvin JE, Uryu K, Lee VM et al. (1999) Axon pathology in Parkinson's disease and Lewy body dementia hippocampus contains alpha-, beta-, and gamma-synuclein. Proc Natl Acad Sci USA 96: 13450-13455.

109. Kanda S, Bishop JF, Eglitis MA et al. (2000) Enhanced vulnerability to oxidative stress by α-synuclein mutations and C-terminal truncation. Neurosci 97: 279-284.
110. Haavik J, Toska K (1998) Tyrosine hydroxylase and Parkinson's disease. Mol Neurobiol 16: 285-309.

UBIQUITIN AND THE MOLECULAR PATHOLOGY OF NEURODEGENERATIVE DISEASES

[1]James Lowe, [2]John Mayer, [2]Michael Landon and [2]Robert Layfield
Schools of [2]Biomedical Sciences and [1]Clinical Laboratory Sciences. University of Nottingham Medical School, UK. james.lowe@nottingham.ac.uk

1. INTRODUCTION

Ubiquitin is an important molecule which now appears central to several pathological cellular processes[1]. There is a close association between ubiquitin-protein conjugates, cytoskeletal abnormalities as well as with the endosome-lysosome system in cells. This overview concentrates on the involvement of the ubiquitin system in diseases of the nervous system, an importance which has been reinforced by recent molecular insights into neurodegenerative processes.

In normal cells the ubiquitin-proteasome system is involved in both the cytosolic and endoplasmic reticulum degradation of proteins. In this system, free ubiquitin is formed into chains and attached to target proteins. In this process ATP-dependent activation of ubiquitin takes place in which it is coupled to an activation enzyme termed E1. Ubiquitin is transferred to a ubiquitin carrier protein termed E2, and then can be coupled to a protein substrate via interaction with a ubiquitin-protein ligase complex termed E3. Poly-ubiquitin chains can be attached to proteins and these are recognised and degraded by a large multiprotein complex termed the 26S proteasome. Ubiquitin is regenerated and goes back to a free pool in the cell by the activity of a deubiquitinating enzyme (ubiquitin-C-terminal hydrolase)[2-4]. There are many different proteins with activities as E1, E2, E3 or

Neuropathology and Genetics of Dementia, Edited by Tolnay and Probst
Kluwer Academic/Plenum Publishers, New York 2001

deubiquitinating activity[1]. For example, genome sequencing has identified around 90 deubiquitinating enzymes, making this group the largest family of enzymes in the ubiquitin system[5]. Ubiquitin is also involved in the endosome-lysosome system but its role in this cell compartment is less clear[6].

Molecules with a structural similarity to ubiquitin have also been identified and are termed Ub-like proteins (Ubls). They are divided into two subclasses:

- type-1 Ubls, e.g. NEDD8, Sentrin/SUMO, and Apg12, that can be attached to target proteins in a manner analogous to that seen in the ubiquitin pathway

- type-2 Ubls that are characterised by the presence of a ubiquitin-like structure within a large protein with an apparently distinct function, e.g. Parkin, Rad23, and Elongin B[7,8]

The biological and neuropathological significance of these proteins and post-translational modifications are just being explored.

Ubiquitin appears to be incorporated into inclusion bodies and protein accumulations characteristic of several of the major neurodegenerative diseases and investigation of the roles of ubiquitin is thus of great importance in extending our understanding of these disorders[9]. These diseases include Alzheimer's disease, Parkinson's disease, Creutzfeldt Jakob disease (prion encephalopathy), amyotrophic lateral sclerosis (motor neuron disease), multiple system atrophy, nuclear hyaline inclusion disease, Huntington's disease and types of spinocerebellar ataxia characterised by intranuclear polyglutamine-containing inclusions.

2. UBIQUITIN IMMUNOHISTOCHEMISTRY IN NEUROPATHOLOGY

Antibodies to ubiquitin are now essential tools for the investigation of neurodegenerative diseases[10,11]. The availability of immunohistochemical probes for ubiquitin, enzymes of the ubiquitination and deubiquitination pathways, and proteasome-related proteins, has made it possible to study the biogenesis of abnormal protein accumulations. The most widely used anti-ubiquitin antisera are rabbit polyclonals raised against ubiquitin conjugated to keyhole limpet haemocyanin[12]. These antibodies have a high affinity for ubiquitin-protein conjugates but are less sensitive in detecting free ubiquitin. Conversely, many monoclonal antibodies raised against synthetic peptides or native ubiquitin have a relatively low affinity for ubiquitin-protein conjugates and a high affinity for free ubiquitin. Antibodies are also available which are specific for ubiquitin-ubiquitin branches in poly-ubiquitin chains[13]. Many antibodies to ubiquitin also detect ubiquitin fusion proteins and those raised against ubiquitin-protein conjugates may detect ubiquitin-cross-reactive protein (UCRP), an interferon-inducible 15kDa protein that has sequence homology with ubiquitin[14,15]. Cross-reactivity with UCRP should be taken into account when interpreting studies using anti-ubiquitin antibodies[16].

PGP9.5 is a deubiquitinating enzyme found at about 2% of total protein in nerves[17-22]. Immunohistochemical studies show that PGP9.5 is found in several ubiquitinated inclusion bodies[23].

Immunoreactivity to ubiquitin-activating enzyme, E1, has been localised to neurofibrillary tangles in hippocampal neurons[6].

Components of the 20S proteasome have been localised to cortical Lewy bodies[24]. An antiserum reactive against a regulatory ATPase (designated subunit 6 or TBP7), immunostains all of the pathological features of Alzheimer's disease that involve abnormal tau protein[6].

3. DISEASE-RELATED STRUCTURES SEEN WITH ANTI-UBIQUITIN ANTIBODIES

A large number of neuronal disease-related structures can be immunostained using antibodies to ubiquitin. These can be classified as follows:

- Accumulation in the endosome/lysosome system
- The family of ubiquitin/intermediate filament/αB crystallin inclusions
- Inclusions based on tau protein accumulation
- Nuclear inclusions
- Ubiquitin/filament inclusions of motor neuron disease
- Cytoplasmic neuritic accumulations

3.1 Ubiquitin staining due to accumulation in the endosome/lysosome System

Ubiquitin dot-like bodies can be seen in both cortex and white matter in the normal ageing brain in a variety of species. Increased density of such bodies is seen in association with a variety of degenerative diseases[25,26]. Ultrastructural studies have shown that dot-like bodies correspond to two types of dystrophic neurite; one results from the dilatation of nerve axons that accumulate lysosomal dense bodies, while the other is caused by myelin degeneration[25,27].

Ubiquitin immunostaining is seen in plaque neurites in Alzheimer's disease in which the presence of ubiquitin, as well as APP, can be seen in lysosome-related dense bodies[28-31]. Ubiquitin-immunoreactive neurites are also seen associated with amyloid plaques composed of Aß, in the absence of tau protein, as seen in normal ageing[34] and dementia with Lewy bodies[30]. The accumulation of APP and ubiquitin in abnormal axonal processes is not restricted to Alzheimer's disease[28]. Any possible pathogenic role of ubiquitin in this situation is uncertain. Aß has been suggested to interfere with the proteolytic activity of proteasomes[35] which may explain why ubiquitinated tau protein accumulates within neurons in Alzheimer's disease. Ubiquitin has been suggested to have a role in the internalisation of endosomal membranes[36] and its presence in dense bodies in dystrophic neurites may be related to increased endocytosis. Ubiquitin has also been suggested to act as a cofactor that enhances amyloid formation[37-39], and if this is the case the secretion of ubiquitin with Aß from the lysosome-related dense bodies in nerve terminals may have a direct causative role in amyloid deposition.

Lysosome-related ubiquitin immunostaining is seen in brain tissue in transmitted, sporadic and genetic forms of prion disease[40,41]. This is a feature seen very early in the course of incubation in transmitted disease in mice[42-44].

3.2 The intermediate filament/αB crystallin/ubiquitin family

The intermediate filament/αB crystallin/ubiquitin inclusions represent a family of related inclusion bodies based on the common features of aggregation of a protein substrate together with intermediate filaments in association with ubiquitin, αB crystallin, certain enzymes of the ubiquitin-dependent proteolytic pathway and proteasomes[45-50]. The inclusions that fall into this category include Lewy bodies, Rosenthal fibres, cytoplasmic bodies, and Mallory's hyaline seen in hepatocytes (Table 1). For some of the diseases characterised by this type of inclusion, a mutant or aberrant inclusion-specific protein has been found to be the cause of inclusion biogenesis. This has been best exemplified by the involvement of alpha synuclein in Lewy bodies, seen in cases caused by mutation in the alpha synuclein gene as well as sporadic cases[51-53]. Similarly, Mallory's hyaline can be produced in transgenic animals due to lack of expression of keratin 18[54]. This is an interesting model as inclusions only develop in aged mice, strengthening links with the age-related neurodegenerative diseases. Embracing the concept of this family of inclusions suggests that mutations in any of the proteins demonstrated in sporadic instances of this type of inclusion might also explain less common genetic causes of disease. This seems to be the case with mutations in alpha synuclein, PGP9.5, αB crystallin and intermediate filaments[55-59]. We have previously suggested that this type of inclusion is the morphological manifestation of a protein-eliminating stress response of the cell[48]. We have proposed that the formation of ubiquitin/intermediate filament/αB crystallin inclusions is probably a common cellular response to chronic sub-lethal injury and is likely to be cytoprotective[9,23,45,48,60]. The term aggresome has more recently been proposed to embrace this concept of a structural reorganisation of the cytoskeleton in response to abnormal protein[61].

Name	Main associated disease	Filament Protein	Cell type
Lewy body	Parkinson's disease	Neurofilament protein	Neuron
Cortical Lewy body	Dementia with Lewy bodies	Neurofilament protein	Neuron
Rosenthal fibre	Alexander's disease	Glial fibrillary acidic protein	Astrocyte
Cytoplasmic body	Myopathy	Desmin	Skeletal muscle cell
Mallory body	Alcoholic liver disease	Cytokeratin	Hepatocyte
Crook's hyaline	Cushing's disease	Cytokeratin	Endocrine cell
Hyaline inclusion	Lichen amyloidosus	Cytokeratin	Skin epithelium cell

Table 1. Inclusions based on aggregates of intermediate filaments

3.3 Tau-protein inclusions

The tau-containing neurofibrillary tangles of Alzheimer's disease are frequently ubiquitinated[62-64] with the suggestion that ubiquitination is a late event[65]. Tangles in other neurodegenerative conditions are in our experience only faintly and inconsistently ubiquitin immunoreactive. In Alzheimer's disease there is also widespread accumulation of tau protein associated with ubiquitin in dilated nerve cell processes in the form of neuropil threads[66]. Immunohistochemistry has shown that the ends of threads are devoid of ubiquitin immunoreactivity, suggesting that ubiquitination of tau protein may be a late phenomenon[67]. Protein chemical studies in Alzheimer's disease have shown that the ubiquitin conjugation sites are localised to the microtubule-binding region of tau, which is predominantly mono-ubiquitinated[68]. Personal experience has shown considerable case-to-case variation in the extent of tangle ubiquitination in Alzheimer's disease, a phenomenon that is presently unexplained but is most likely to be related to polymorphisms in the genes coding form elements of the complex ubiquitin-proteasome degradative pathway.

Pick bodies are another tau-containing neuronal inclusion that also show anti-ubiquitin immunoreactivity[45,69,70], although usually in a faint or granular pattern compared to other inclusions. In contrast to the family of intermediate filament/αB crystallin-containing inclusions, the vast majority of neurofibrillary tangles and Pick bodies do not contain the ubiquitin C-terminal hydrolase PGP9.5[23,49].

Glial cytoplasmic inclusions of multiple system atrophy (MSA) contain tau protein[71-73], and additionally contain ubiquitinated αB crystallin[74] as well as alpha synuclein[71]. The absence of intermediate filaments separates this type of inclusion from the intermediate filament/αB crystallin/ubiquitin family of inclusions. Although MSA is now widely regarded as a synucleinopathy, together with Parkinson's disease, it is important to consider that the main defect in MSA may not lie with alpha synuclein but may lie elsewhere in a more complex interplay between the constituent proteins demonstrated in these distinctive inclusions.

3.4 Nuclear filamentous inclusions

Inclusions in nuclei are now a recognised hallmark of the family of inherited trinucleotide expansion disorders characterised by CAG repeats coding for polyglutamine tracts in the disease protein. Such nuclear inclusions containing polyglutamines can be seen in neurons in cases of human disease as well as in neurons from transgenic mouse models of this class of disorder. Inclusions are filamentous and show immunoreactivity for the disease protein as well as ubiquitin. Antibodies that detect the polyglutamine tracts also show immunoreactivity in these inclusions[75-84]. There is evidence to suggest proteasome involvement in the biogenesis of inclusions[85]. Such proteins may also accumulate in neurites in addition to neuronal nuclei with the suggestion that there may be more than one pathway to neuronal death in such conditions[86].

Marinesco bodies are ubiquitinated filamentous inclusions seen in nuclei in association with ageing[25]. Similar inclusions are observed in the rare condition nuclear hyaline inclusion disease[87-89]. The molecular nature of the filaments in these inclusions remains undetermined but there are reports of immunoreactivity for antibodies that detect polyglutamine. Ultrastructural appearances suggests that they might be based on intermediate filaments; hence a candidate protein would be one of the nuclear lamins.

3.5 Filamentous inclusions in motor neuron disease and frontotemporal dementia

Filamentous ubiquitin reactive inclusions are a hallmark of both sporadic and familial forms of motor neuron disease (MND)[90-97] and were originally termed skeins by Leigh. These inclusions are seen in motor neurons but have also been found in non-motor areas[98]. The molecular nature of the filamentous inclusions in MND remains uncertain as they have been found to be unreactive to a wide range of antisera to cytoskeletal proteins[99]. Inclusions do not contain PGP9.5[23], do not contain αB crystallin[49], and do not appear to be immunoreactive for proteasomes[24]. Similar inclusions have also been described in hereditary dentatorubral-pallidoluysian atrophy[100].

In a small proportion of familial cases of MND, disease is related to a mutation in the gene coding for superoxide dismutase-1 (SOD1). Immunohistochemical studies have shown that inclusions in such cases may be immunoreactive to SOD1, as well as to ubiquitin and phosphorylated neurofilament protein. This suggests that mutant SOD1 might be interacting with ubiquitin and intermediate filaments in the elimination of this abnormal protein[101]. Transgenic models of SOD1 MND have also shown inclusions with both ubiquitin and SOD1 immunoreactivity[90].

Similar inclusions can also be seen in non-motor systems and have been linked with several clinical syndromes:

- Motor neuron disease with frontotemporal dementia (sporadic and familial)[102-104]
- Pure frontotemporal dementia or semantic dementia[102,104-109]
- Clinical corticobasal degeneration[110]
- Primary progressive aphasia[111-112]
- Multisystem degeneration[113]

In some cases, ubiquitin inclusions in neurons are also seen in association with ubiquitin-immunoreactive neurites, seen in areas of microvacuolation in outer cortical areas[105,108,109]. These changes overlap with those described in some cases of frontotemporal dementia where neuronal inclusions are not seen[114].

The relationship between these conditions is presently unclear. On the basis of morphological similarities and well documented families in which MND, MND dementia and pure frontotemporal dementia can be seen, a relationship with MND has been suggested for this group of conditions. The name motor neuron disease-inclusion dementia has been coined in recognition that such patients have the inclusions but not the clinical features of motor neuron disease[105]. Other authors have taken the view that there is insufficient evidence to relate such cases to MND. There is clearly a pressing need to characterise the protein within the inclusions and neurites in these conditions. Based on past precedents in other conditions the protein found in sporadic cases will give the gene to explain some familial cases. If the link with MND is correct then this gene will also explain another subgroup of familial MND.

3.6 Ubiquitin accumulation in the cytosol of nerve cell processes

Several neurodegenerative diseases have been shown to be characterised by dilated nerve cell processes in the cerebral cortex immunoreactive for ubiquitin but some being also immunoreactive for other disease-specific proteins. There are two explanations for ubiquitin immunoreactivity in axonal swellings. In certain forms of axonal pathology, notably where the axon becomes ballooned and there is continued axonal transport, ubiquitin immunoreactivity appears to be related to the accumulation of lysosome-related dense bodies. Following traumatic severance of nerves and in some neurodegenerative diseases, axons develop swelling of their proximal terminal portion to form spheroids. Such spheroids contain microtubules, abundant neurofilament protein, synaptic vesicles, mitochondria and numerous lysosome-related bodies that accumulate in the dilated end as a result of impaired axonal transport systems. Immunohistochemical studies indicate that spheroids are ubiquitin immunoreactive, mainly from ubiquitin in lysosome-related vesicles[115-116]. Such swellings also exhibit high levels of PGP9.5.

In dementia with Lewy bodies, Huntington's disease and some type of frontotemporal dementia associated with ubiquitin inclusions, dilated neuronal processes become filled with ubiquitin reactive material which appears filamentous. In Lewy body disorders and in Huntington's disease the accumulation is based on disease-associated proteins alpha synuclein[117-120] and huntingtin[88,121,122]. In MND one would predict that the neurites also contain a specific disease-associated protein, yet to be discovered. Last in this section, axonal spheroids are also a feature of rare inherited degenerative diseases of the nervous system, ubiquitin immunoreactivity being seen in these structures in both brain iron accumulation type 1 (Hallervorden-Spatz disease) and infantile neuroaxonal dystrophy[123]. As such axonal swellings also show alpha synuclein immunoreactivity[124,125] with Lewy bodies being a frequent associated feature in brain iron accumulation type 1, this suggests that this class of abnormality may be best related to the intermediate filament/ubiquitin family of disorders.

4. MECHANISTIC LINKS BETWEEN THE UBIQUITIN SYSTEM AND INCLUSION FORMATION

There are several possible mechanistic links between the ubiquitin system and neurodegenerative disease

Firstly, an abnormal protein substrate may be present in cells and although tagged by ubiquitin for degradation this accumulates, possibly by slower degradation kinetics than for the wild-type protein. It has been suggested that this may explain the accumulation of alpha synuclein[126] and may be a potential factor in tau protein, huntingtin and other polyglutamine-containing accumulations in neurons[127]. Chaperone proteins that normally assist in the degradative process may not be able to interact with the abnormal conformation of mutant proteins[128]. Alternatively, cell stress proteins with chaperone activity may bind to the misfolded protein and prevent ubiquitin-mediated degradation[129].

Secondly, there may be mutations or polymorphisms in key component of the ubiquitin system leading to inefficiency in the degradative pathway, leading to disease. This has been suggested for certain types of Parkinson's disease (PD). Parkin is an E3 ubiquitin ligase[130] and loss of function mutations in its gene are linked to one familial form of PD[131]. Mutation in the gene for PGP9.5, a neuronal ubiquitin-C-hydrolase[21] has been linked to a rare form of familial PD[58,132,133] with suggestion that polymorphisms may affect the risk of developing sporadic PD[134]. Mice with a mutation in the neuronal ubiquitin-C-hydrolase develop a phenotype of neuroaxonal dystrophy termed gracile axonal dystrophy[135-137]. The idea that the ubiquitin system is working in a cytoprotective fashion is strengthened by data which shows that while mutation of the E6-AP ubiquitin ligase reduces the frequency of nuclear inclusions in SCA1 mice this inefficiency of the ubiquitin system causes more severe neuronal damage[127].

Thirdly, ubiquitinated material may accumulate because inhibition of the proteasome system leads to inefficiency in the degradative pathway. Inhibition of the chymotrypsin-like activity of the proteasome results in accumulation of ubiquitinated proteins in cells[138] and it seems logical that some disorders may be the result of inhibition of the proteasome system. Such insights are now starting to be being addressed through transgenic models, but no conclusive information is yet available. There are indications from studies in yeast that defects in proteasome subunits may result in accumulation of ubiquitin-protein conjugates. In understanding the pathogenesis of Alzheimer's disease experiments *in vitro* suggest that Aß can decrease ubiquitin-dependent protein degradation by selectively inhibiting the chymotrypsin-like activity of the proteasome, raising the possibility that such inhibition *in vivo* could cause the intracellular accumulation of ubiquitinated proteins[35]. Frameshift mutant ubiquitin seen in Alzheimer's disease and other age-related accumulations of tau protein[139-141], has been suggested to inhibit the proteasome[142]. In a transgenic model of polyglutamine inclusions, inhibition of proteasome ubiquitin-dependent degradation increased the formation of intranuclear aggregates, lending support for a relationship between the activity of this part of the ubiquitin degradative pathway and formation of inclusions[143]. Accumulation of ubiquitinated proteins as a consequence of proteasome inhibition has also been suggested as a stimulus for pro-inflammatory responses[144].

5. SUMMARY

Ubiquitin plays a central role in normal cellular function as well as in disease. It is possible to group ubiquitin-immunostained structures into several main groups, the most distinctive being the ubiquitin/intermediate filament/αB crystallin family of inclusions that seem to represent a general cellular response to abnormal proteins recently termed the aggresomal response. While ubiquitin immunohistochemistry is a very useful technique for detecting pathological changes and inclusion bodies in the nervous system this alone is not enough to classify inclusions, and a panel of antibodies is recommended to clarify any findings made by screening tissues with anti-ubiquitin.

Several mechanistic possibilities now exist to explain the accumulation of ubiquitinated proteins in cells of the nervous system, understanding of which should lead to new therapeutic advances in the group of chronic neurodegenerative diseases.

REFERENCES

1. Peters J, Harris J, Finley D (1998) Ubiquitin and the biology of the cell. Plenum Press, New York and London.
2. Ciechanover A, Orian A, Schwartz AL (2000) Ubiquitin-mediated proteolysis: biological regulation via destruction. Bioessays 22: 442-451.
3. Ciechanover A, Orian A, Schwartz AL (2000) The ubiquitin-mediated proteolytic pathway: mode of action and clinical implications. J Cell Biochem 34 (Supplement): 40-51.
4. Kornitzer D, Ciechanover A (2000) Modes of regulation of ubiquitin-mediated protein degradation. J Cell Physiol 182: 1-11.
5. Chung CH, Baek SH (1999) Deubiquitinating enzymes: their diversity and emerging roles. Biochemical & Biophysical Research Communications 266: 633-640.

6. Mayer R, Landon M, Lowe J (1998). Ubiquitin and the molecular pathology of human disease. In: Ubiquitin and the biology of the cell. (Peters J, Harris J, Finley D, eds), pp 429-462: Plenum Press, New York and London.
7. Tanaka K, Suzuki T, Chiba T (1998) The ligation systems for ubiquitin and ubiquitin-like proteins. Molecules & Cells 8: 503-512.
8. Yeh ET, Gong L, Kamitani T (2000) Ubiquitin-like proteins: new wines in new bottles. Gene 248: 1-14.
9. Mayer RJ, Arnold J, Laszlo L et al. (1991) Ubiquitin in health and disease. Biochim Biophys Acta 1089: 141-157.
10. Lowe J (1998) Establishing a pathological diagnosis in degenerative dementias. Brain Pathol 8: 403-406.
11. Munoz DG (1999) Stains for the differential diagnosis of degenerative dementias. Biotechnic & Histochemistry 74: 311-320.
12. Haas AL, Bright PM (1985) The immunochemical detection and quantitation of intracellular ubiquitin-protein conjugates. J Biol Chem 260: 12464-12473.
13. Fujimuro M, Sawada H, Yokosawa H (1994) Production and characterization of monoclonal antibodies specific to multi-ubiquitin chains of polyubiquitinated proteins. FEBS Lett 349: 173-180.
14. Ahrens PB, Besancon F, Memet S et al. (1990) Tumour necrosis factor enhances induction by beta-interferon of a ubiquitin cross-reactive protein. J Gen Virol 71: 1675-1682.
15. Loeb KR, Haas AL (1992) The interferon-inducible 15-kDa ubiquitin homolog conjugates to intracellular proteins. J Biol Chem 267: 7806-7813.
16. Lowe J, McDermott H, Loeb K et al. (1995) Immunohistochemical localization of ubiquitin cross-reactive protein in human tissues. J Pathol 177: 163-169.
17. Day IN, Hinks LJ, Thompson RJ (1990) The structure of the human gene encoding protein gene product 9.5 (PGP9.5), a neuron-specific ubiquitin C-terminal hydrolase. J Biochem 268: 521-524.
18. Ermisch B, Schwecheimer K (1995) Protein gene product (PGP) 9.5 in diagnostic (neuro-) oncology. An immunomorphological study. Clin Neuropathol 14: 130-136.
19. Harris MD, Moore IE, Steart PV et al. (1990) Protein gene product (PGP) 9.5 as a reliable marker in primitive neuroectodermal tumours-an immunohistochemical study of 21 childhood cases. Histopathology 16: 271-277.
20. Thompson RJ, Doran JF, Jackson P et al. (1983) PGP 9.5-a new marker for vertebrate neurons and neuroendocrine cells. Brain Res 278: 224-228.
21. Wilkinson KD, Lee KM, Deshpande S et al. (1989) The neuron-specific protein PGP 9.5 is a ubiquitin carboxyl-terminal hydrolase. Science 246: 670-673.
22. Wilson PO, Barber PC, Hamid QA (1988) The immunolocalization of protein gene product 9.5 using rabbit polyclonal and mouse monoclonal antibodies. Br J Exp Pathol 69: 91-104.
23. Lowe J, McDermott H, Landon M et al. (1990) Ubiquitin carboxyl-terminal hydrolase (PGP 9.5) is selectively present in ubiquitinated inclusion bodies characteristic of human neurodegenerative diseases. J Pathol 161: 153-160.
24. Ii K, Ito H, Tanaka K et al. (1997) Immunocytochemical co-localization of the proteasome in ubiquitinated structures in neurodegenerative diseases and the elderly. J Neuropathol Exp Neurol 56: 125-131.
25. Dickson DW, Wertkin A, Kress Y et al. (1990) Ubiquitin immunoreactive structures in normal human brains. Distribution and developmental aspects. Lab Invest 63: 87-99.

26. Pappolla MA, Omar R, Saran B (1989) The 'normal' brain. 'Abnormal' ubiquitinilated deposits highlight an age-related protein change. Am J Pathol 135: 585-591.
27. Migheli A, Attanasio A, Pezzulo T et al. (1992) Age-related ubiquitin deposits in dystrophic neurites: an immunoelectron microscopic study. Neuropathol Appl Neurobiol 18: 3-11.
28. Cochran E, Bacci B, Chen Y et al. (1991) Amyloid precursor protein and ubiquitin immunoreactivity in dystrophic axons is not unique to Alzheimer's disease. Am J Pathol 139: 485-489.
29. Cras P, Kawai M, Lowery D et al. (1991) Senile plaque neurites in Alzheimer's disease accumulate amyloid precursor protein. Proc Natl Acad Sci USA 88: 7552-7556.
30. Dickson DW, Crystal H, Mattiace LA et al. (1989) Diffuse Lewy body disease: Light and electron microscopic immunocytochemistry of senile plaques. Acta Neuropathol 78: 572-584.
31. Dickson DW, Crystal HA, Mattiace LA et al. (1992) Identification of normal and pathological aging in prospectively studied nondemented elderly humans. Neurobiol Aging 13: 179-189.
32. Dickson DW, Wertkin A, Mattiace LA et al. (1990) Ubiquitin immunoelectron microscopy of dystrophic neurites in cerebellar senile plaques of Alzheimer's disease. Acta Neuropathol 79: 486-493.
33. Yasuhara O, Kawamata T, Aimi Y et al. (1994) Two types of dystrophic neurites in senile plaques of Alzheimer disease and elderly non-demented cases. Neurosci Lett 171: 73-76.
34. He Y, Delaere P, Duyckaerts C et al. (1993) Two distinct ubiquitin immunoreactive senile plaques in Alzheimer's disease: relationship with the intellectual status in 29 cases. Acta Neuropathol 86: 109-116.
35. Gregori L, Fuchs C, Figueiredo-Pereira ME et al, (1995) Amyloid beta-protein inhibits ubiquitin-dependent protein degradation in vitro. J Biol Chem 270: 19702-19708.
36. Hicke L, Riezman H (1996) Ubiquitination of a yeast plasma membrane receptor signals its ligand-stimulated endocytosis. Cell 84: 277-287.
37. Alizadeh-Khiavi K, Normand J, Chronopoulos S et al. (1991) Alzheimer's disease brain-derived ubiquitin has amyloid-enhancing factor activity: Behavior of ubiquitin during accelerated amyloidogenesis. Acta Neuropathol 81: 280-286.
38. Alizadeh-Khiavi K, Normand J, Chronopoulos S et al. (1992) Amyloid enhancing factor activity is associated with ubiquitin. Virchows Arch of Pathol 420: 139-148.
39. Chronopoulos S, Lembo P, Alizadeh-Khiavi K et al. (1992) Ubiquitin: its potential significance in murine AA amyloidogenesis. J Pathol 167: 249-259.
40. Ironside JW, McCardle L, Hayward PA et al. (1993) Ubiquitin immunocytochemistry in human spongiform encephalopathies. Neuropathol Appl Neurobiol 19: 134-140.
41. Migheli A, Attanasio A, Vigliani MC et al. (1991) Dystrophic neurites around amyloid plaques of human patients with Gerstmann-Straussler-Scheinker disease contain ubiquitinated inclusions. Neurosci Lett 121: 55-58.
42. Laszlo L, Lowe J, Self T et al. (1992) Lysosomes as key organelles in the pathogenesis of prion encephalopathies. J Pathol 166: 333-341.
43. Lowe J, Fergusson J, Kenward N et al. (1992) Immunoreactivity to ubiquitin-protein conjugates is present early in the disease process in the brains of scrapie-infected mice. J Pathol 168: 169-177.
44. Lowe J, McDermott H, Kenward N et al. (1990) Ubiquitin conjugate immunoreactivity in the brains of scrapie infected mice. J Pathol 162: 61-66.

45. Lowe J, Blanchard A, Morrell K et al. (1988) Ubiquitin is a common factor in intermediate filament inclusion bodies of diverse type in man, including those of Parkinson's disease, Pick's disease, and Alzheimer's disease, as well as Rosenthal fibres in cerebellar astrocytomas, cytoplasmic bodies in muscle, and Mallory bodies in alcoholic liver disease. J Pathol 155: 9-15.
46. Lowe J, Mayer RJ (1989) Ubiquitin: New insights into chronic degenerative diseases. Br J Hosp Med 42: 462-466.
47. Lowe J, Mayer RJ (1990) Ubiquitin, cell stress and diseases of the nervous system. Neuropathol Appl Neurobiol 16: 281-291.
48. Lowe J, Mayer RJ, Landon M (1993) Ubiquitin in neurodegenerative diseases. Brain Pathol 3: 55-65.
49. Lowe J, McDermott H, Pike I et al. (1992) alphaB crystallin expression in non-lenticular tissues and selective presence in ubiquitinated inclusion bodies in human disease. J Pathol 166: 61-68.
50. Manetto V, Abdul-Karim FW, Perry G et al. (1989) Selective presence of ubiquitin in intracellular inclusions. Am J Pathol 134: 505-513.
51. Goedert M (1999) Filamentous nerve cell inclusions in neurodegenerative diseases: tauopathies and alpha-synucleinopathies. Philosophical Transactions of the Royal Society of London - Series B: Biological Sciences 354: 1101-1118.
52. Solano SM, Miller DW, Augood SJ et al. (2000) Expression of alpha-synuclein, parkin, and ubiquitin carboxy-terminal hydrolase L1 mRNA in human brain: genes associated with familial Parkinson's disease. Ann Neurol 47: 201-210.
53. Trojanowski JQ, Lee VM (1998) Aggregation of neurofilament and alpha-synuclein proteins in Lewy bodies: implications for the pathogenesis of Parkinson disease and Lewy body dementia. Arch Neurol 55: 151-152.
54. Magin TM, Schroder R, Leitgeb S et al. (1998) Lessons from keratin 18 knockout mice: formation of novel keratin filaments, secondary loss of keratin 7 and accumulation of liver-specific keratin 8-positive aggregates. J Cell Biol 140: 1441-1451.
55. Bova MP, Yaron O, Huang Q et al. (1999) Mutation R120G in alphaB-crystallin, which is linked to a desmin-related myopathy, results in an irregular structure and defective chaperone-like function. Proc Natl Acad Sci USA 96: 6137-6142.
56. Dalakas MC, Park KY, Semino-Mora C et al. (2000) Desmin myopathy, a skeletal myopathy with cardiomyopathy caused by mutations in the desmin gene. NEJM 342: 770-780.
57. Farrer M, Gwinn-Hardy K, Hutton M et al. (1999) The genetics of disorders with synuclein pathology and parkinsonism. Hum Mol Genet 8: 1901-1905.
58. Leroy E, Boyer R, Auburger G et al. (1998) The ubiquitin pathway in Parkinson's disease. Nature 395: 451-452.
59. Vicart P, Caron A, Guicheney P et al. (1998) A missense mutation in the alphaB-crystallin chaperone gene causes a desmin-related myopathy. Nat Genet 20: 92-95.
60. Mayer RJ, Lowe J, Lennox G et al. (1989) Intermediate filament-ubiquitin diseases: implications for cell sanitization. Biochem Soc Symp 55: 193-201.
61. Johnston JA, Ward CL, Kopito RR (1998) Aggresomes: a cellular response to misfolded proteins. J Cell Biol 143: 1883-1898.
62. Cole GM, Timiras PS (1987) Ubiquitin-protein conjugates in Alzheimer's lesions. Neurosci Lett 79: 207-212.
63. Mori H, Kondo J, Ihara Y (1987) Ubiquitin is a component of paired helical filaments in Alzheimer's disease. Science 235: 1641-1644.

64. Perry G, Friedman R, Shaw G (1987) Ubiquitin is detected in neurofibrillary tangles and senile plaque neurites of Alzheimer disease brains. Proc Natl Acad Sci USA 84: 3033-3036.
65. Bancher C, Brunner C, Lassmann H et al. (1989) Accumulation of abnormally phosphorylated tau precedes the formation of neurofibrillary tangles in Alzheimer's disease. Brain Res 477: 90-99.
66. Perry G, Kawai M, Tabaton M (1991) Neuropil threads of Alzheimer's disease show a marked alteration of the normal cytoskeleton. J Neurosci 11: 1748-1755.
67. Iwatsubo T, Hasegawa M, Esaki Y et al. (1992) Lack of ubiquitin immunoreactivities at both ends of neuropil threads. Possible bidirectional growth of neuropil threads. Am J Pathol 140: 277-282.
68. Morishima M, Ihara Y (1994) Posttranslational modifications of tau in paired helical filaments. Dementia 5: 282-288.
69. Love S, Saitoh TQS, Cole GM (1988) Alz-50, ubiquitin and tau immunoreactivity of neurofibrillary tangles, Pick bodies and Lewy bodies. J Neuropathol Exp Neurol 47: 393-405.
70. Murayama S, Mori H, Ihara Y et al. (1990) Immunocytochemical and ultrastructural studies of Pick's disease. Ann Neurol 27: 394-405.
71. Dickson DW, Lin W, Liu WK et al. (1999) Multiple system atrophy: a sporadic synucleinopathy. Brain Pathol 9: 721-732.
72. Papp MI, Kahn JE, Lantos PL (1989) Glial cytoplasmic inclusions in the CNS of patients with multiple system atrophy (striatonigral degeneration, olivopontocerebellar atrophy and Shy-Drager syndrome). J Neurol Sci 94: 79-100.
73. Papp MI, Lantos PL (1992) Accumulation of tubular structures in oligodendroglial and neuronal cells as the basic alteration in multiple system atrophy. J Neurol Sci 107: 172-82.
74. Tamaoka A, Mizusawa H, Mori H et al. (1995) Ubiquitinated alphaB-crystallin in glial cytoplasmic inclusions from the brain of a patient with multiple system atrophy. J Neurol Sci 129: 192-198.
75. Bates GP, Mangiarini L, Davies SW (1998) Transgenic mice in the study of polyglutamine repeat expansion diseases. Brain Pathol 8: 699-714.
76. Becher MW, Kotzuk JA, Sharp AH et al. (1998) Intranuclear neuronal inclusions in Huntington's disease and dentatorubral and pallidoluysian atrophy: correlation between the density of inclusions and IT15 CAG triplet repeat length. Neurobiol Disease 4: 387-397.
77. Davies SW, Turmaine M, Cozens BA et al. (1997) Formation of neuronal intranuclear inclusions underlies the neurological dysfunction in mice transgenic for the HD mutation. Cell 90: 537-548.
78. DiFiglia M, Sapp E, Chase KO et al. (1997) Aggregation of huntingtin in neuronal intranuclear inclusions and dystrophic neurites in brain. Science 277: 1990-1993.
79. Duyckaerts C, Durr A, Cancel G et al. (1999) Nuclear inclusions in spinocerebellar ataxia type 1. Acta Neuropathol 97: 201-207.
80. Gourfinkel-An I, Cancel G, Duyckaerts C et al. (1998) Neuronal distribution of intranuclear inclusions in Huntington's disease with adult onset. Neuroreport 9: 1823-1826.
81. Hayashi Y, Kakita A, Yamada M et al. (1998) Hereditary dentatorubral-pallidoluysian atrophy: detection of widespread ubiquitinated neuronal and glial intranuclear inclusions in the brain. Acta Neuropathol 96: 547-552.

82. Koyano S, Uchihara T, Fujigasaki H et al. (1999) Neuronal intranuclear inclusions in spinocerebellar ataxia type 2: triple-labeling immunofluorescent study. Neurosci Lett 273: 117-20.
83. Schmidt T, Landwehrmeyer GB, Schmitt I et al. (1998) An isoform of ataxin-3 accumulates in the nucleus of neuronal cells in affected brain regions of SCA3 patients. Brain Pathology 8: 669-679.
84. Sieradzan KA, Mechan AO, Jones L et al. (1999) Huntington's disease intranuclear inclusions contain truncated, ubiquitinated huntingtin protein. Exp Neurol 156: 92-99.
85. Chai Y, Koppenhafer SL, Shoesmith SJ et al. (1999) Evidence for proteasome involvement in polyglutamine disease: localization to nuclear inclusions in SCA3/MJD and suppression of polyglutamine aggregation in vitro. Hum Mol Genet 8: 673-682.
86. Aronin N, Kim M, Laforet G et al. (1999) Are there multiple pathways in the pathogenesis of Huntington's disease? Philosophical Transactions of the Royal Society of London - Series B: Biological Sciences 354: 995-1003.
87. Bergmann M, Kuchelmeister K, Kryne KB et al. (1994) Infantile multiple system atrophy with cytoplasmic and intranuclear glioneuronal inclusions. Acta Neuropathol 87: 642-647.
88. Kimber TE, Blumbergs PC, Rice JP et al. (1998) Familial neuronal intranuclear inclusion disease with ubiquitin positive inclusions. J Neurol Sci 160: 33-40.
89. Takahashi J, Fukuda T, Tanaka J et al. (2000) Neuronal intranuclear hyaline inclusion disease with polyglutamine-immunoreactive inclusions. Acta Neuropathol 99: 589-594.
90. Ince PG, Lowe J, Shaw PJ (1998) Amyotrophic lateral sclerosis: current issues in classification, pathogenesis and molecular pathology. Neuropathol Appl Neurobiol 24: 104-117.
91. Leigh PN, Anderton BH, Dodson A et al. (1988) Ubiquitin deposits in anterior horn cells in motor neurone disease. Neurosci Lett 93: 187-203.
92. Lowe J (1994) New pathological findings in amyotrophic lateral sclerosis. J Neurol Sci 124: 38-51.
93. Lowe J, Aldridge F, Lennox G et al. (1989) Inclusion bodies in motor cortex and brainstem of patients with motor neurone disease are detected by immunocytochemical localisation of ubiquitin. Neurosci Lett 105: 3-7.
94. Lowe J, Lennox G, Jefferson D et al. (1988) A filamentous inclusion body within anterior horn neurones in motor neurone disease defined by immunocytochemical localisation of ubiquitin. Neurosci Lett 94: 203-210.
95. Migheli A, Autilio Gambetti L, Gambetti P et al. (1990) Ubiquitinated filamentous inclusions in spinal cord of patients with motor neuron disease. Neurosci Lett 114: 5-10.
96. Murayama S, Mori H, Ihara Y et al. (1990) Immunocytochemical and ultrastructural studies of lower motor neurons in amyotrophic lateral sclerosis. Ann Neurol 27: 137-148.
97. Schiffer D, Autilio Gambetti L, Chio A et al. (1991) Ubiquitin in motor neuron disease: Study at the light and electron microscope. J Neuropathol Exp Neurol 50: 463-473.
98. Anderson VE, Cairns NJ, Leigh PN (1995) Involvement of the amygdala, dentate and hippocampus in motor neuron disease. J Neurol Sci 129: 75-78.
99. Mather K, Martin JE, Swash M et al. (1993) Histochemical and immunocytochemical study of ubiquitinated neuronal inclusions in amyotrophic lateral sclerosis. Neuropathol Appl Neurobiol 19: 141-145.

100. Hayashi Y, Kakita A, Yamada M et al. (1998) Hereditary dentatorubral-pallidoluysian atrophy: ubiquitinated filamentous inclusions in the cerebellar dentate nucleus neurons. Acta Neuropathol 95: 479-482.
101. Shibata N, Hirano A, Kobayashi M et al. (1996) Intense superoxide dismutase-1 immunoreactivity in intracytoplasmic hyaline inclusions of familial amyotrophic lateral sclerosis with posterior column involvement. J Neuropathol Exp Neurol 55: 481-490.
102. Cooper PN, Jackson M, Lennox G et al. (1995) Tau, ubiquitin, and alphaB-crystallin immunohistochemistry define the principal causes of degenerative frontotemporal dementia. Arch Neurol 52: 1011-1015.
103. Kawashima T, Kikuchi H, Takita M et al. (1998) Skein-like inclusions in the neostriatum from a case of amyotrophic lateral sclerosis with dementia. Acta Neuropathol 96: 541-545.
104. Mann DM (1998) Dementia of frontal type and dementias with subcortical gliosis. Brain Pathol 8: 325-338.
105. Jackson M, Lennox G, Lowe J (1996) Motor neurone disease-inclusion dementia. Neurodegeneration 5: 339-350.
106. Jackson M, Lowe J (1996) The new neuropathology of degenerative frontotemporal dementias. Acta Neuropathol 91: 127-134.
107. Kertesz A, Kawarai T, Rogaeva E et al. (2000) Familial frontotemporal dementia with ubiquitin-positive, tau-negative inclusions. Neurology 54: 818-827.
108. Kinoshita A, Tomimoto H, Suenaga T et al. (1997) Ubiquitin-related cytoskeletal abnormality in frontotemporal dementia: immunohistochemical and immunoelectron microscope studies. Acta Neuropathol 94: 67-72.
109. Rossor MN, Revesz T, Lantos PL et al. (2000) Semantic dementia with ubiquitin-positive tau-negative inclusion bodies. Brain 123: 267-276.
110. Grimes DA, Bergeron CB, Lang AE (1999) Motor neuron disease-inclusion dementia presenting as cortical-basal ganglionic degeneration. Mov Disorders 14: 674-680.
111. Kertesz A, Munoz DG (1997) Primary progressive aphasia. Clin Neurosci 4: 95-102.
112. Tsuchiya K, Ozawa E, Fukushima J et al. (2000) Rapidly progressive aphasia and motor neuron disease: a clinical, radiological, and pathological study of an autopsy case with circumscribed lobar atrophy. Acta Neuropathol 99: 81-87.
113. Bergmann M, Kuchelmeister K, Migheli A (1993) Motor neuron disease with pallido-luysio-nigral atrophy. Acta Neuropathol 86: 105-108.
114. Tolnay M, Probst A (1995) Frontal lobe degeneration: Novel ubiquitin-immunoreactive neurites within frontotemporal cortex. Neuropathol Appl Neurobiol 21: 492-497.
115. Li GL, Farooque M (1996) Expression of ubiquitin-like immunoreactivity in axons after compression trauma to rat spinal cord. Acta Neuropathol 91: 155-160.
116. Schweitzer JB, Park MR, Einhaus SL et al. (1993) Ubiquitin marks the reactive swellings of diffuse axonal injury. Acta Neuropathol 85: 503-507.
117. Dickson DW, Ruan D, Crystal H et al. (1991) Hippocampal degeneration differentiates diffuse Lewy body disease (DLBD) from Alzheimer's disease: Light and electron microscopic immunocytochemistry of CA2-3 neurites specific to DLBD. Neurology 41: 1402-1409.
118. Dickson DW, Schmidt ML, Lee VM et al. (1994) Immunoreactivity profile of hippocampal CA2/3 neurites in diffuse Lewy body disease. Acta Neuropathol 87: 269-276.
119. Gai WP, Blessing WW, Blumbergs PC (1995) Ubiquitin-positive degenerating neurites in the brainstem in Parkinson's disease. Brain 118: 1447-1459.

120 Kim H, Gearing M, Mirra SS (1995) Ubiquitin-positive CA2/3 neurites in hippocampus coexist with cortical Lewy bodies. Neurology 45: 1768-1770.
121. Cammarata S, Caponnetto C, Tabaton M (1993) Ubiquitin-reactive neurites in cerebral cortex of subjects with Huntington's chorea: a pathological correlate of dementia? Neurosci Lett 156: 96-98.
122. Jackson M, Gentleman S, Lennox G et al. (1995) The cortical neuritic pathology of Huntington's disease. Neuropathol Appl Neurobiol 21: 18-26.
123. Malandrini A, Cavallaro T, Fabrizi GM et al. (1995) Ultrastructure and immunoreactivity of dystrophic axons indicate a different pathogenesis of Hallervorden-Spatz disease and infantile neuroaxonal dystrophy. Virchows Arch of Pathology 427: 415-421.
124. Arawaka S, Saito Y, Murayama S et al. (1998) Lewy body in neurodegeneration with brain iron accumulation type 1 is immunoreactive for alpha-synuclein. Neurology 51: 887-9.
125. Newell KL, Boyer P, Gomez-Tortosa E et al. (1999) Alpha-synuclein immunoreactivity is present in axonal swellings in neuroaxonal dystrophy and acute traumatic brain injury. J Neuropathol Exp Neurol 58: 1263-1268.
126. Bennett MC, Bishop JF, Leng Y et al. (1999) Degradation of alpha-synuclein by proteasome. J Biol Chem 274: 33855-33858.
127. Cummings CJ, Reinstein E, Sun Y et al. (1999) Mutation of the E6-AP ubiquitin ligase reduces nuclear inclusion frequency while accelerating polyglutamine-induced pathology in SCA1 mice. Neuron 24: 879-892.
128. Cummings CJ, Mancini MA, Antalffy B et al. (1998) Chaperone suppression of aggregation and altered subcellular proteasome localization imply protein misfolding in SCA1. Nat Genet 19: 148-154.
129. Wyttenbach A, Carmichael J, Swartz J et al. (2000) Effects of heat shock, heat shock protein 40 (HDJ-2), and proteasome inhibition on protein aggregation in cellular models of Huntington's disease. Proc Natl Acad Sci USA 97: 2898-2903.
130. Shimura H, Hattori N, Kubo S et al. (2000) Familial Parkinson disease gene product, parkin, is a ubiquitin-protein ligase. Nat Genet 25: 302-305.
131. Kitada T, Asakawa S, Hattori N et al. (1998) Mutations in the parkin gene cause autosomal recessive juvenile parkinsonism. Nature 392: 605-608.
132. Harhangi BS, Farrer MJ, Lincoln S et al. (1999) The Ile93Met mutation in the ubiquitin carboxy-terminal-hydrolase-L1 gene is not observed in European cases with familial Parkinson's disease. Neurosci Lett 270: 1-4.
133. Lincoln S, Vaughan J, Wood N et al. (1999) Low frequency of pathogenic mutations in the ubiquitin carboxy-terminal hydrolase gene in familial Parkinson's disease. Neuroreport 10: 427-429.
134. Maraganore DM, Farrer MJ, Hardy JA et al. (1999) Case-control study of the ubiquitin carboxy-terminal hydrolase L1 gene in Parkinson's disease. Neurology 53: 1858-1860.
135. Mukoyama M, Yamazaki K, Kikuchi T et al. (1989) Neuropathology of gracile axonal dystrophy (GAD) mouse. An animal model of central distal axonopathy in primary sensory neurons. Acta Neuropathol 79: 294-299.
136. Saigoh K, Wang YL, Suh JG et al. (1999) Intragenic deletion in the gene encoding ubiquitin carboxy-terminal hydrolase in gad mice. Nat Genet 23: 47-51.
137. Yamazaki K, Wakasugi N, Tomita T et al. (1988) Gracile axonal dystrophy (GAD), a new neurological mutant in the mouse. Proc Soc Exp Biol Med 187: 209-215.

138. Figueiredo PM, Berg KA, Wilk S (1994) A new inhibitor of the chymotrypsin-like activity of the multicatalytic proteinase complex (20S proteasome) induces accumulation of ubiquitin-protein conjugates in a neuronal cell. J Neurochem 63: 1578-1581.
139. Fergusson J, Landon M, Lowe J et al. (2000) Neurofibrillary tangles in progressive supranuclear palsy brains exhibit immunoreactivity to frameshift mutant ubiquitin-B protein. Neurosci Lett 279: 69-72.
140. van Leeuwen FW, Fischer DF, Benne R et al. (2000) Molecular misreading. A new type of transcript mutation in gerontology. Ann NY Acad Sci 908: 267-281.
141. van Leeuwen FW, Hol EM (1999) Molecular misreading of genes in Down syndrome as a model for the Alzheimer type of neurodegeneration. J Neural Transm Supplementum: 137-159.
142. Lam Y, Pickart C, Alban A et al. (2000) Inhibition of the ubiquitin-proteasome system in Alzheimer's disease. Proc Natl Acad Sci USA 97: 9902-9906.
143. de Cristofaro T, Affaitati A, Cariello L et al. (1999) The length of polyglutamine tract, its level of expression, the rate of degradation, and the transglutaminase activity influence the formation of intracellular aggregates. Biochemical & Biophysical Research Communications 260: 150-158.
144. Rockwell P, Yuan H, Magnusson R et al. (2000) Proteasome inhibition in neuronal cells induces a proinflammatory response manifested by upregulation of cyclooxygenase-2, its accumulation as ubiquitin conjugates, and production of the prostaglandin PGE(2). Arch Biochem Biophys 374: 325-333.

PERISOMATIC GRANULES OF HIPPOCAMPAL CA1 NEURONS IN ALZHEIMER'S DISEASE, PRE-ALZHEIMER STAGE AND PICK'S DISEASE: AN OVERLOOKED PATHOLOGICAL ENTITY

Alphonse Probst, Claudia Mistl, Sabine Ipsen and Markus Tolnay
Institute of Pathology, Division of Neuropathology, Basel University, Basel, Switzerland

1. INTRODUCTION

There is overall consensus that neuronal loss occurs selectively in vulnerable areas of the brain in Alzheimer's disease (AD) compared to age-matched normal specimens. Significant and disease specific loss of hippocampal neurons is known to occur in CA1 neuronal subsets which appear to be more severely affected than other hippocampal subregions[1]. However, the cause of cell death in the hippocampus remains unknown. One possibility is that, in some neurons at least, the accumulation of tau protein in an aggregated fibrillar form (the neurofibrillary tangle-NFT) results in the demise of the cell, although the way tau filaments interfere with neuronal function is largely unknown. Other proposed mechanisms leading to cell death are oxidative stress or excitatory amino acid (EAA) receptor-mediated exitotoxicity[2].

Alzheimer's and age-related pathology of the hippocampal pyramidal cells mainly affect the CA1 sector and encompass a series of well established pathological changes. These include early hyperphosphorylation and somatodendritic localisation of abnormally phosphorylated tau protein[3], the formation of neurofibrillary tangles and dendritic threads, the granulovacuolar degeneration (GVD) of neuronal cytoplasma[4], the formation of rod-like intraneuronal inclusions, the Hirano bodies, and spindle-shaped tau-immunoreactive dilatations of apical dendrites in low Braak stages[5]. In more severe AD cases, variable numbers of extraneuronal tangles (ghost tangles) can be seen in CA1 as a direct evidence of NFT-associated cell death in this cortex area[3].

Looking at ubiquitin-stained sections of the hippocampus routinely, we have repeatedly observed the presence of small ubiquitin-reactive neuropil granules immediately adjacent to the soma of CA1 pyramidal cells in subjects with AD and in some non-demented old subjects with Alzheimer type changes in their hippocampus. No equivalent structures could be found in H&E stained sections of AD cases, even when carefully looking at the perisomatic area of pyramidal cells. A search for similarly looking structures in reports on ubiquitin pathology in the hippocampus was unsuccessful. However, our attention was attracted by a recent report on the distribution of glutamate receptors (GluR) in the hippocampus of normal and AD patients[6]. This study describes a peculiar type of „non-plaque" dystrophic neurites in the perisomatic neuropil of CA1 pyramidal cells. In addition to labelling neuronal cytoplasma, antibodies to AMPA GluR1 and GluR2 revealed juxta-neuronal clusters of granular reactivity reminiscent of the above mentioned ubiquitin-positive perisomatic granules (PSG).

PSG have been undoubtedly overlooked in the past which may be explained by the fact that PSG are ubiquitin-stained, a property shared by most GVD granules. Furthermore, the location of both types of granules near or in close contact to the cytoplasmic membrane of neurons, makes their distinction difficult. Both GVD and PSG are intracellular but one is apparently located outside the cell soma (the PSG), the other (the GVD bodies) inside the soma of CA1 neurons.

We have therefore been interested in comparing PSG as revealed by antibodies to ubiquitin and AMPA GluR subtypes in order to find out if they correspond to one and the same structure. GVD will be described in some detail in order to be compared with PSG. In further sections PSG will be discussed in the context of hippocampal cell pathology and connectivity. The nosologic specificity of PSG will be evaluated by including other neurodegenerative disorders like argyrophilic grain disease and Pick's disease.

2. FEATURES OF PERISOMATIC GRANULES

PSG, as shown with ubiquitin antibodies (Figure 1) and specific antibodies to AMPA GluR subtype 1 (GluR1) (the rabbit anti-rat AMPA GluR1 antibody cross-reacts with human GluR1; Upstate Biotechnology) and GluR2 (mouse monoclonal antibody to human AMPA receptor GluR2;

Novocastra) on paraffin sections, consisted of 1 to 4μm wide immunoreactive granules, distributed around the somata of CA1 pyramidal cells.

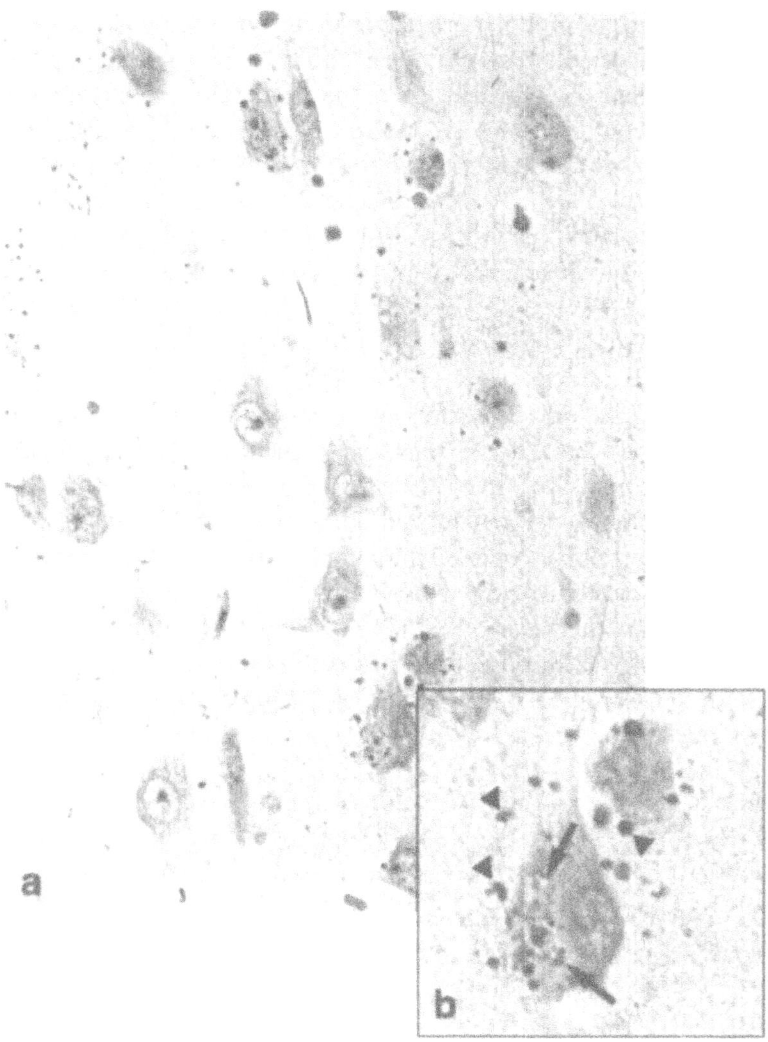

Figure 1. **a.** Perisomatic granules and GVD in hippocampal CA1 sector in an AD patient. Immunostaining with ubiquitin antibody. **b.** Inset: some perisomatic granules are shown by arrowheads and GVD granules by arrows. Note that some GVD granules fail to stain for ubiquitin.

In semithin sections immunostained according to modified Lane and Europa method[7] (epon was removed by alcoholic sodium hydroxyde and bleached by an aqueous solution) PSG could be more precisely localized than in paraffin sections and were seen to be located outside the neuronal soma. Some PSG were in close contact with the soma of neurons, while others were found as isolated granules at some distance from neuronal cell bodies or in close contact with initial segments of basal or apical dendrites. PSG were not found along more distal dendritic segments in the stratum radiatum as shown by dual ubiquitin and MAP2 labelling on the same section. PSG failed to stain using various antibodies to phosphorylated or non-phosphorylated epitopes of tau, PGP8, SMI32, SMI31, antibodies against APP, neurofilament polypeptides, alphaB-crystallin and alpha-synuclein.

More importantly, ubiquitin stained PSG and AMPA GluR2 immunostained granules were found to have similar size and shape. Both were also found to largely coincide in double GluR2 and ubiquitin immunolabelled paraffin sections.

Immunoelectronmicroscopy using AMPA GluR2 antibody revealed enlarged (dystrophic) strongly immunoreactive boutons near neuronal cell bodies in CA1 neuropil, some measuring up to 5μm in diameter. Strong GluR2 reactivity was seen along the cell membrane and the synaptic region i.e. the contact regions with unstained boutons, probably presynaptic axonal teminals. The immunostained boutons were filled with tubulovesicular or floccular material, with some tubular structures in close contact with the synaptic region or the plasma membrane. The lesions were not associated with amyloid fibrils and did not contain paired helical filaments. In contrast to previous findings by Aronica et al.[6], no Hirano body filaments could be found in close association with PSG.

3. THE GRANULOVACUOLAR DEGENERATION

GVD refer to the presence of a conspicuous structure consisting of a dense granule (0.5 to 1.5μm)in vacuole-like clear area about twice the diameter of the granules found typically in the cytoplasm of hippocampal pyramidal cells[4]. The affected cells are usually found in the CA1 sector but, upon careful examination of the brain, GVD can also be seen in other parts of the hippocampus and even outside the allocortex (see below)[8]. The degree of granulovacuolar change only slightly increases during normal brain aging. In AD, however, the „granulovacuolar index“ exceeds by many times that of age-matched controls[9]. The GVD is not specific for AD or for the hippocampus, although its association with both is characteristic and well established[10,11]. In fact, in AD, large number of cells containing GVD also contain NFT[12].

The work of Okamoto and coworkers[13] (1991), based upon electron microscopic observation of well preserved autopsy tissue, suggested that GVD might be derived by autophagy. Depending upon the local constituents of the cytoplasm, autophagic vacuoles might be predicted to be heterogeneous. In fact, immunocytochemical studies have shown the presence of a variety of proteins in GVD among them ubiquitin seems the most consistent[14].

The physiological importance of this observation is unclear since ubiquitin is a key component of the proteasome-mediated proteolysis unrelated to the autophagic lysosomal pathway thought to underlie GVD body formation[15]. A process of cytoplasmic sequestration is involved in autophagy where macromolecules and organelles are subject to a process of non selective volume uptake of the cell cytoplasma[15]. This may account for the inconsistent presence of ubiquitin in GVD bodies. As GVD shows significant association with „pretangle" pathology of tau protein (see below), the occasional occurence of hyperphosphorylated tau in GVD granules is also likely. However, the presence of tau in GVD granules remains a controversial issue. Some monoclonal antibodies raised against PHF-tau where shown to label GVD granules[16], however, it has not been possible to identify tau or neurofilaments directly in these lesions[17]. One explanation for the staining results obtained with monoclonal antibodies to phosphorylared tau and neurofilament epitopes is that these antigens may cross-react wih phosphoepitopes found within unrelated phosphoproteins[15]. In contrast, immunoreactivity for casein-kinase 1 isoforms, which is not mediated by phosphoepitopes, was found to be dramatically increased in GVD granules and in NFT in AD brains[15]. Recently, activated caspase-3 has been detected within the matrix of GVD granules[18]. Caspase-3 reactivity was restricted to the GVD granules and not present in other cytoplasmic compartments of the corresponding neurons. The activation of caspase-3 in GVD was interpreted as a self-repair attempt of the affected neuron leading to the sequestration in autophagic vacuoles of cytoplasmic compartments where the potentially dangerous activation of proapoptotic enzymes has occured.

The amount of GVD per cell has been correlated with the severity of dementia and with the extent of neuronal loss[9,19,20]. One frequent observation in severe AD cases is that of enlarged CA1 pyramidal cells entirely filled with vacuoles and granules of GVD, without the presence of a neurofibrillary tangle in the same cell. This suggests that GVD *per se*, may eventually be associated with cell death when its role in cell repair is exhausted.

Variable numbers of GVD were observed among our AD cases and in some old non-demented patients with only very few GVD found in individuals with Braak stages I-II. GVD was mainly found in cells of CA1 sector but some were also encountered in the subiculum and in other parts of the brain. GVD granules measured approximately 0.5 to 1.5μm in diameter and were only inconsistently labelled with ubiquitin antibodies. In semithin sections immunostained for ubiquitin, GVD granules disclosed an inhomogenous structure and appeared fragmented. Some strongly stained granules were surrounded by an ubiquitin stained membrane and by an abundant amorphous ubiquitin positive intracytoplasmic material closely apposed to the outer membrane of GVD vacuoles. Most granules of GVD strongly reacted with SMI31 antibody. Even in cells entirely filled with GVD, the SMI31-reactive granules were completely confined within the limits of the neuronal soma.

4. PSG ARE DISTINCT FROM GVD

The main differences between PSG and GVD are listed in Table 1. In sections stained for ubiquitin, PSG were of larger average diameter (1 to 4 μm) than granules of GVD (0.5 to 1.5μm). Most granules of GVD strongly reacted with SMI31 antibody which was not the case for extrasomatic PSG. This was confirmed in semithin sections where extracellular and intracellular granules could be more clearly separated. In addition, GVD granules remained consistently unstained with GluR1 or GluR2 antibodies. The distribution of neurons with GVD was undoubtedly more extent than that of PSG. Most GVD were found in CA1 sector but some were also found in CA2-3 and the subiculum and other parts of the brain, whereas PSG were entirely confined to the CA1 sector and subiculum.

	PSG	GVD
Distribution	CA1 (subiculum)	CA1 prosubiculum, CA2, CA3-4
Relationship to neuronal soma	perisomatic	intrasomatic
Size	1-4μm	0.5-1.5μm (GVD granule!)
Immunocytochemistry:		
Ubiquitin	positive	variable positive
AMPA GluR 1-2	positive	negative
SMI 31	negative	positive

Table 1. PSG and GVD: main distinguishing features.

Although most GVD disclosed ubiquitin staining, in some neurons, a large proportion of granules remained unstained in paraffin as well as in plastic embedded sections. The largest immunostained granules disclosed an inhomogenous structure and appeared fragmented whereas PSG showed more homogenous ubiquitin staining.

5. PSG ARE ASSOCIATED WITH PYRAMIDAL (PRINCIPAL) CA1 CELLS AND NOT WITH INTERNEURONS

PSG are found in close vicinity to neurons of the CA1 sector, the question therefore arises around which neuronal subtype they occur. Neurons of different types, among them the pyramidal or principal cells and the non-pyramidal cells or interneurons, are known to occur in the hippocampal formation. Several types of non-pyramidal cells have been reported in the various layers of the CA1 sector of hippocampus by using glutamic acid decarboxylase (GAD)[21] or parvalbumin immunolabelling[22]. The somata of GAD positive cells of the CA1 sector offer a wide range of shape but are generally multiform to fusiform with the long axis aligned parallel or vertically to the alveus[21]. PSG are obviously located around pyramidal-shaped neuronal somata, thus interneurons do not appear a likely candidate for PSG. This was also confirmed by dual immunolabelling with ubiquitin and parvalbumin antibodies. Ubiquitin-reactive PSG were almost exclusively found in association with parvalbumine negative (pyramidal) cells and were absent from the surface of parvalbumine-reactive interneurons.

6. RELATIONSHIP OF PSG AND OTHER PATHOLOGIES OF CA1 PYRAMIDAL CELLS

There was an obvious relationship between the numbers of CA1 neurons surrounded by PSG and the extent of neurofibrillary changes in the hippocampus according to Braak stages[23]. Whereas no or only very scarce PSG could be found in cognitively unimpaired patients with early transentorhinal Braak stages I-II, and even in some individuals with limbic stage III, the density of pyramidal cells with PSG was definitely increased in limbic stage IV and in isocortical stages V and VI.

Looking at double ubiquitin/tau immunolabelled sections, it was obvious that, with few exceptions, PSG occured in relation to pyramidal cells whose soma and dendrites presented diffuse immunoreactivity for hyperphosphorylated tau protein, a staining pattern corresponding to „pretangle“ stage of pathology of pyramidal cells. For instance, staining with monoclonal anti-tau antibody AT8[24] of nerve cells surrounded by PSG was of variable intensity with higher number of PSG found around intensely labelled neurons.

As shown by combined ubiquitin/Gallyas stain, PSG did generally not occur in relation to neurons containing a neurofibrillary tangles. Only exceptionally some isolated PSG were seen near a neuron containing a Gallyas-positive tangle. In such cells, the Gallyas-positive inclusion also disclosed ubiquitin reactivity. In contrast to PSG, GVD were often associated with NFT. Furthermore, GVD was much more often found in AT8-immunostained neurons than in AT8-negative cells. A significant association between neurons with AT8 positive cytoskeletal alterations, i.e. neurofibrillary tangles, pretangle changes, and GVD was recently established by Stadelmann et al.[18].

7. PSG ARE NOT SPECIFIC FOR AD

To test if PSG are AD specific changes or if they occur in neurodegenerative disorders other than AD, we also examined hippocampi from subjects with argyrophilic grain disease (AgD) and Pick's disease. AgD cases were all characterized by great numbers of argyrophilic Gallyas and tau-positive (phosphodependent monoclonal antibodies were used), but only in part ubiquitin-positive, neuropil grains throughout the hippocampus, the entorhinal cortex and the amygdala. In the hippocampus, the argyrophilic grains (ArGs) were chiefly found in CA1 sector and the prosubiculum. ArGs were associated with moderate number of neurofibrillary tangles whose distribution corresponded to early Braak stages I-III and with large numbers of tau immunoreactive „pretangle neurons“[25,26]. Few ubiquitin positive perisomatic clusters could be recognized in the CA1 sector in AgD cases, but PSG may have been overlooked since ArGs, like PSG, are ubiquitin positive. In Pick's disease, in contrast, large perisomatic ubiquitin positive granules were easily found close to CA1 pyramidal cells. As most Pick bodies disclosed some degree of ubiquitin reactivity, it was possible to determine if pyramidal cells surrounded by PSG contained a Pick body. In fact, about 80% of cells surrounded by PSG were containing a Pick body and 20% were not. In addition, a variable proportion of pyramidal cells, with or without PSG or Pick body, also disclosed GVD. Interestingly, only few PSG per neuronal soma were found inspite of quite a large number of neurons surrounded by PSG.

8. GENERAL DISCUSSION

PSG, as revealed by GluR1-2 and ubiquitin-immunohistochemistry, are a consistent feature of the CA1 stratum pyramidum in AD but can already be found in some subjects with Braak stage III. PSG are not AD specific as they also form part of the hippocampal pathology in other „tauopathies" like Pick's disease and AgD.

PSG, like punctate granular GluR1 and 2 reactivity described by Aronica and coworkers[6] in the hippocampus, are restricted to CA1 sector and are in their majority juxtaneuronal and almost exclusively perisomatic with almost no PSG found in the stratum radiatum i.e. along apical processes of pyramidal cells.

At the ultrastructural level, immunoreactivity for AMPA GluR was found to correspond to cellular processes, whose size and localization near neuronal somata were compatible with PSG seen by light microscopy. The contact with smaller unstained boutons containing synaptic vesicles, suggested that the GluR reactive cellular processes could correspond to postsynaptic structures. The content of these lesions was similar to that described in Aronica's account, and consisted of a densely immunoreactive tubulovesicular or floccular material without evidence for the presence of paired helical filaments.

PSG represent a novel type of ubiquitin reactive lesion in the diseased brain. They were found to be distinct from other ubiquitin reactive changs known to occur in the AD context, including neuropil threads or dystrophic neurites of plaque periphery. They were also different from dot-like age-related ubiquitin deposits in dystrophic neurites, which are mostly localized in the white matter and correspond to myelinated dystrophic neurites filled with ubiquitin-labelled lysosomal dense bodies[27]. In conclusion, our findings confirm PSG as a new and hitherto unrecognized entity, identical to the juxtacellular AMPA GluR1 and 2 positive neuropil grains described by Aronica and coworkers[6] and different from GVD, which are intrasomatic, AMPA GluR negative and of smaller average size (see Table 1).

PSG are detected in a specific region of the hippocampus (the CA1 sector) which is especially vulnerable to neurodegeneration in AD and in some other „tauopathies" like AgD[28]. In particular, the number of PSG in CA1 was found to increase from stage III to VI of Braak together with increasing numbers of GVD. Our findings also suggest a pathogenic link between PSG and tau protein pathology in hippocampal pyramidal cells, mainly hyperphosphorylation of tau protein in the so called „pretangle" neurons. On the other side, no or only sparse PSG were found around nerve cell bodies containing neurofibrillary tangles. In AD cases, PSG were not associated with „pretangle" neurons in other cortical areas than CA1 sector.

In Pick's disease, the hippocampal formation generally displays dramatic atrophy and high densities of Pick bodies, particularly in the dentate gyrus[29]. However, like in AD, PSG were only found in the CA1 and not in the more intensely affected dentate gyrus or in other parts of the brain, thus pointing to a strong regional succeptibility for PSG.

The mechanisms underlying CA1 succeptibility for PSG and the reasons for the formation of PSG close to the plasma membrane of CA1 principal cell somata are unclear. Also largely unexplained is the selective AMPA GluR involvement. The NMDA or kainate receptor subunits were not found to be involved in PSG[6]. Possible explanations for the abnormal GluR and ubiquitin immunoreactivity around CA1 neurons in AD and in other tauopathies could include increased expression of AMPA receptors in dystrophic or degenerating dendrites or an altered AMPA GluR compartmentalization and ubiquitination in degenerating hippocampal neurons[6]. One possibility is that glutamate receptor-mediated excitotoxicity leads to cytoskeletal alterations in discrete domains of pyramidal neurons, leading to the focal neuritic changes we described as PSG.

One may speculate if increased AMPA expression is occuring at postsynaptic sites corresponding to glutamatergic projections from other principal CA1 neurons or collaterals from principal cells, although excitatory dendritic interactions among CA1 principal cells is relatively sparse[30] and glutamatergic contacts mainly occur on basal dendrites of principal cells. Contacts on the soma of principal cells mainly originate from GABAergic and parvalbumine positive basket cells and are therefore hardly a candidate for postsynaptic GluR immunoreactive PSG. GluR2 regulatory subunit of AMPA receptor family was also localized on the soma and dendrites of GABAergic CA1 interneurons[31]. PSG however, were exclusively localized near the surface of principal cell somata making this possibility unlikely. In summary, explaining PSG in terms of neuronal connectivity in the CA1 subfield is presently unsatisfactory. PSG clearly needs to be investigated with improved anatomical resolution bearing in mind that they may represent an important link in pathogenic events contributing to neuronal death in the hippocampus.

REFERENCES

1. West MJ, Coleman PD, Flood DG et al. (1994) Differences in the pattern of hippocampal neuronal loss in normal ageing and Alzheimer's disease. Lancet 344: 769-772.
2. Papolla MA, Omar RA, Kim KS et al. (1992) Immunohistochemical evidence of oxidative stress in Alzheimer's disease. Am J Pathol 140: 621-628.

3. Braak E, Braak H, Mandelkow EM (1994) A sequence of cytoskeletal changes related to the formation of neurofibrillary tangles and neuropil threads. Acta Neuropathol 87: 554-567.
4. Simchowicz T (1911) Histologische Studien über die senile Demenz. Histologische und histopathologische Arbeiten über die Grosshirnrinde 4: 267-444.
5. Braak E, Braak H (1997) Alzheimer's disease: transiently developing dendritic changes in pyramidal cells of sector CA1 of the Ammon's horn. Acta Neuropathol 93: 323-325.
6. Aronica E, Dickson DW, Kress Y et al. (1998) Non-plaque dystrophic dendrites in Alzheimer's disease hippocampus: A new pathological structure revealed by glutamate receptor immunocytochemistry. Neurosci 82: 979-991.
7. Lane BP, Europa DL (1967) Differential staining of ultrathin sections of epon-embedded tissues for light microscopy. J Histochem Cytochem 13: 579-582.
8. Xu M, Shibayama H, Kobayashi H et al. (1992) Granulovacuolar degeneration in the hippocampal cortex of aging and demented patients; a quantitative study. Acta Neuropathol 85 : 1-9.
9. Ball MJ, Lo P (1977) Granulovacuolar degeneration in aging brain and in dementia. J Neuropathol Exp Neurol 36: 474-487.
10. Woodard JS (1962) Clinicopathological significance of granulovacuolar degeneration in Alzheimer's disease. J Neuropathol Exp Neurol 21: 85-91.
11. Tomlinson BE, Blessed G, Roth M (1968) Observations on the brain of non demented old people. J Neurol Sci 7: 331-356.
12. Grüntha E (1930) Die pathologische Anatomie der senilen Demenz und der Alzheimerschen Krankheit. In: Handbuch der Geisteskrankheiten. Die Anatomie der Psychosen (Bumke O ed), pp638-672, Vol. 11: Springer Verlag, Berlin.
13. Okamoto K, Hirai S, Lizuka T et al. (1991) Re-examination of granulovacuolar degeneration. Acta Neuropathol 82: 340-345.
14. Love S, Saitoh T, Quijada S et al. (1988) Ubiquitin and tau reactivity of neurofibrillary tangles, Pick bodies and Lewy bodies. J Neuropathol Exp Neurol 47: 3939-4005.
15. Goshal N, Smiley JF, De Maggio AJ et al. (1999) A new link between the fibrillar and granulovacuolar lesions of Alzheimer's disease. Am J Pathol 155: 1163-1171.
16. Dickson DW, Liu WK, Kress Y et al. (1993) Phosphorylated tau immunoreactivity of granulovacular degeneration (GVD) of Alzheimer's disease: localization of two amino terminal tau epitopes in GVD. Acta Neuropathol 85: 463-470.
17. Kahn J, Anderton BH, Probst A (1985) Immunohistochemical study of granulo-vacuolar degeneration using monoclonal antibodies to neurofilaments. J Neurol Neurosurg Psychiatry 48: 924-926.
18. Stadelmann C, Deckwerth TL, Srinivasan A et al. (1999) Activation of caspase-3 in single neurons and autophagic granules of granulovacuolar degeneration in Alzheimer's disease. Evidence for apoptotic cell death. Am J Pathol 155: 1459-1466.
19. Ball MJ (1978) Topographic distribution of neurofibrillary tangles and granulovacuolar degeneration in hippocampal cortex of aging and demented patients: a quantitative study. Acta Neuropathol 42: 73-80.
20. Tomlinson BE, Kitchener D (1972) Granulovacuolar degeneration of hippocampal pyramidal cells. J Pathol 106: 165-185.
21. Braak E, Olbrich HG, Braak H et al. (1986) Glutamic acid decarboxylase immunoreactivity in sector CA1 of the human hippocampus. Anat Embryol 175: 15-23.
22. Braak E, Strotkamp B, Braak H (1991) Parvalbumin-immunoreactive structures in the hippocampus of the human adult. Cell Tissue Res 264: 33-48.

23. Braak H, Braak E (1991) Neuropathological stageing of Alzheimer-related changes. Acta Neuropathol 82: 239-259.
24. Goedert M, Jakes R, Vanmechelen E (1995) Monoclonal antibody AT8 recognizes tau protein phosphorylated at both serine 202 and threonine 205. Neurosci Lett 189: 167-170.
25. Tolnay M, Mistl C, Ipsen S et al. (1998) Argyrophilic grains of Braak: occurrence in dendrites of neurons containing hyperphosphorylated tau protein. Neuropathol Appl Neurobiol 24: 53-59.
26. Tolnay M, Spillantini MG, Goedert M et al. (1997) Argyrophilic grain disease: widespread hyperphosphorylation of tau protein in limbic neurons. Acta Neuropathol 93: 477-484.
27. Migheli A, Attanasio A, Gulotta F et al. (1992) Age-related deposits in dystrophic neurites: an immunoelectron microscopic study. Neuropathol Appl Neurobiol 18: 3-11.
28. Tolnay M, Probst A (1999) Review: Tau protein pathology in Alzheimer's disease and related disorders. Neuropathol Appl Neurobiol 25: 171-187.
29. Ball MJ (1979) Topography of Pick inclusion bodies in hippocampi of demented patients. A quantitative study. J Neuropathol Exp Neurol 38: 614-620.
30. Lorente de No R (1934) Studies on the structure of the cerebral cortex- II. Continuation of the study of the ammonic system. J Psychol Neurol 46: 113-177.
31. He Y, Janssen WG, Vissavajjhala P et al. (1998) Synaptic distribution of GluR2 in hippocampal GABAergic interneurons and pyramidal cells: a double-label immunogold analysis. Exp Neurol 150: 1-13.

PATHOLOGICAL MECHANISMS IN POLYGLUTAMINE EXPANSION DISEASES

Didier Devys, Gaël Yvert, Astrid Lunkes, Yvon Trottier, and [1]Jean-Louis Mandel
Institut de Génétique et Biologie Moléculaire et Cellulaire, INSERM/CNRS/Université Louis Pasteur, BP 163, 67 404 Illkirch Cedex, CU de Strasbourg and [1]Hopitaux Universitaires de Strasbourg, France

1. INTRODUCTION

Since 1991, 9 monogenic neurodegenerative diseases have been shown to be caused by moderate expansion of a CAG repeat coding for a polyglutamine stretch in specific target proteins. These disorders include Huntington's disease (HD), spinobulbar muscular atrophy (SBMA) and various spinocerebellar ataxias (SCAs), and their pathogenic mechanism has been the object of intense recent studies. While it is clear that the mutations confer a gain of toxic property to the target proteins, correlated with appearance of a common pathologic epitope and of self aggregation properties, much remains to be learned concerning how these elongated polyglutamines cause neuronal dysfunction and death, and what features can account for the selectivity with which various neuronal populations are affected. Important observations have recently been made, notably by use of cellular or transgenic animal models, that give new insight in the mechanisms of polyglutamine diseases and raise hopes for therapies.

2. POLYGLUTAMINE EXPANSION IN NEURODEGENERATIVE DISORDERS

The class of neurodegenerative diseases caused by an expansion of a CAG/polyglutamine repeat has grown steadily since 1991. The last to be added to the list is a type of spinocerebellar ataxia, named SCA7, that is the only of these diseases that also affects retina[1]. More recently, a single case of polyglutamine expansion arising as a new mutation (from 39 gln to 63 gln) affecting the TATA-binding protein (TBP, a basal transcription factor), has been described in a patient with very severe cerebellar atrophy[2].

All known polyglutamine expansion diseases are progressive disorders that typically begin in adulthood leading to death 10 to 30 years after onset. These neurodegenerative disorders are dominantly inherited except SBMA which is an X-linked disease. An expansion of a CAG repeat in the coding region of the respective genes is the exclusive genetic defect causing HD and other polyglutamine expansion diseases[3]. In all these diseases there is a strong inverse correlation between age of onset of clinical symptoms and the length of the abnormal repeat[4]. The pathological threshold is strikingly similar in HD and four of these diseases (from 35 to 40 gln, Table 1). Below this threshold, the high degree of polymorphism for the CAG/gln repeat observed in the normal population indicates that these proteins tolerate a wide variation in polyglutamine length without adverse effect. The CAG repeats show both somatic and germline instability. However, large expansions are mainly seen upon paternal transmission, a consequence of the greater instability observed in sperm accounting for the paternal bias of anticipation and transmission of the juvenile forms. The 9 diseases are characterized by slowly progressing neuronal degeneration in selected regions of the brain, that differ (with some overlap), between the various diseases (Table 1). This selectivity of neuronal death does not show obvious correlation with the rather ubiquitous pattern of expression of the cognate proteins, with the exception of SCA6[5]. The function of these proteins is unknown in 6 of the 9 cases, the exceptions being the androgen receptor implicated in spinobulbar muscular atrophy (SBMA or Kennedy disease), a calcium channel subunit mutated in SCA6, and the single case of mutation in TBP. The proteins do not resemble each other, except for the polyglutamine tract.

Several arguments are in favor of a "gain of function" conferred by the mutation. The rare HD homozygous patients are not more severely affected than their heterozygote siblings. In contrast, dominant diseases caused by a loss of function mutation are much more severe in the homozygous state[4]. The expansion mutation in SBMA, the first identified polyglutamine expansion disease, causes a gain of toxic property, and not a loss of function of the cognate gene or protein. Mutations that cause a complete loss of function of the androgen receptor gene are responsible for the testicular feminization syndrome, with no neurological symptomatology. By analogy, the same gain of toxic property is expected also to occur in the other diseases, with the possible exception of SCA6. The latter disease is characterized by a much lower pathological threshold of polyglutamine size, and, most importantly, conventional loss of function mutations in the same target gene result in acetazolamide responsive episodic ataxia, a disease that shares overlapping clinical features with SCA6, and that leads to progressive cerebellar degeneration in some patients[5].

Disease	Site of neuropathology	Repeat number		Intracellular localization of encoded protein
		Normal	Disease	
SBMA	Motor neurons (anterior horn cells, bulbar neurons and dorsal root ganglia)	11-34	40-62	nuclear (androgen receptor)
DRPLA	Globus pallidus, dentatorubral and subthlamic nucleus	7-35	49-88	cytoplasmic
HD	Striatum (medium spiny neurons) and cortex in late stage	6-35	36-121	cytoplasmic
SCA1	Cerebellar cortex (Purkinje cells), dentate nucleus and brain stem	6-39	40-81	nuclear, cytoplasmic
SCA2	Cerebellum, pontine nuclei, substantia nigra	15-29	35-64	cytoplasmic
SCA3	Substantia nigra, globus pallidus, pontine nucleus, cerebellar cortex	13-42	61-84	cytoplasmic
SCA6	Cerebellar and mild brain stem atrophy	4-18	21-30	transmembrane (calcium channel subunit α1A)
SCA7	Photoreceptors and bipolar cells, cerebellar cortex, brain stem	7-17	37-300	nuclear, cytoplasmic
"TBP"	Severe cerebellar atrophy	25-42	63 (single case)	nucler (basal transcription factor)

Table 1. Polyglutamine disorders. SBMA; spinobulbar muscular atrophy, DRPLA; dentatorubral-pallidoluysian atrophy, SCA; spinocerebellar ataxia, TBP; TATA-binding protein

3. POLYGLUTAMINE EXPANSION CONFERS NOVEL *IN VITRO* PROPERTIES TO THE TARGET PROTEIN

The finding that the abnormal proteins present in patients with HD, SCA1, SCA2, SCA3 and SCA7 are selectively recognized by a monoclonal antibody (1C2) provided additional support for a common pathogenetic mechanism, implicating an abnormal conformation of elongated polyglutamines (polygln)[6]. This mAb detects huntingtin, by western blot analysis of whole cell protein extracts, above a threshold of about 33-38 gln, and its affinity increases with polygln length. Furthermore, truncated proteins containing expanded polygln also show in vitro a tendency to aggregate into insoluble material, forming rather homogeneous amyloid like fibrils[7,8]. It had been previously proposed that polyglutamine tracts can oligomerize and form ß pleated sheets[9]. The kinetic of aggregation is dependent on polygln length, time, concentration, and is accelerated by seeding with preformed fibrils[10]. It also depends on the nature of the protein, as it was shown that a GST fusion with the N-terminal region of mutated huntingtin aggregates only after cleavage of the GST moiety[8]. The striking similarity of polygln threshold of recognition by 1C2 and of aggregation properties, with the pathological threshold around 36-40 gln in 5 diseases, suggest that these properties reflect features important for pathogenesis.

4. NUCLEAR INCLUSIONS AND MECHANISMS OF NEURODEGENERATION

Attention was called upon nuclear inclusions in polyglutamine diseases, from observations on a mouse model for HD (R6 lines), where expression of a very short truncated huntingtin fragment carrying 115-155 gln (a length corresponding to a disease onset in the first years of life) was leading to a strong neurological phenotype and significant brain shrinkage[11]. Davies et al. reported the conspicuous presence in some but not all areas of the brain of these mice, of nuclear inclusions (NIs) stained with a anti-huntingtin antibody and showing also ubiquitin immunoreactivity[12]. In some regions corresponding to those affected in Huntington's disease, it was estimated that most neurons contain a NI. A similar observation of nuclear inclusions was made in another mouse model, where overexpression, in Purkinje cells, of a full length ataxin 1 containing 82 gln (also corresponding to an infantile onset) leads to severe ataxia, cerebellar dysfunction and ultimately Purkinje cell death[13,14]. In both cases, the appearance of the nuclear inclusions preceded by several weeks the onset of clinical symptoms. Finally, Drosophila models have been created by overexpression of long polygln containing constructs in the developing eye, where intensive eye degeneration is correlated to the formation of NIs[15,16].

Both mouse models have some built in artificial features, such as expression of a truncated protein[11], or gross overexpression of the full length protein[13]. However, similar observations were made in the brains of HD patients, and 5 other diseases (SCA1, 2, 3 and 7, DRPLA and SBMA)[17-20]. In HD, NIs were observed in the cortex and in medium sized neurons of the striatum, but not in non affected brain areas or in controls. These NIs are stained with an antibody specific for the N-terminal region of huntingtin very close to the polyglutamine tract, and by an anti-ubiquitin antibody, but not by antibodies recognizing other regions of huntingtin[18]. In one study the NIs appeared much rarer in adult onset cases and another type of pathological immunoreactive aggregates was observed within dystrophic neurites (distended axon terminals or dendrites)[18]. It was recently proposed that the distribution of dystrophic neurites correlates better with the sites of neurodegeneration than for NIs[21]. It is surprising at first sight that huntingtin or ataxin-3, that are cytoplasmic proteins, appear to aggregate in the nucleus. At least in the case of huntingtin, only a short N-terminal fragment accumulates in the NIs[17,18]. Thus, one may hypothesize that processing of the mutated protein leads to the production of such fragment, that would be transported to the nucleus, where the environment would favor aggregation (Figure 1). Subsequent ubiquitination would reflect a failed attempt to degrade the NIs. This is supported by the general observation that several proteasome subunits and chaperones colocalize with NIs. Recent observations in cellular models indicate that transport to the nucleus is polygln dependent, as short truncated fragments with normal polygln length remain localised to the cytoplasm[22]. It was generally observed in mouse and cellular models that truncated proteins have higher aggregation potential and toxicity than the full-length proteins.

If the presence of NIs is undoubtedly correlated to the diseases, their direct contribution to the toxic process is not yet clarified. Three model systems provide examples of a possible dissociation of polyglutamine toxicity from NIs. Transgenic SCA1 mice were generated in which the transgene construct harboured, in addition to the CAG expansion, a small deletion of a region coding for a self association domain of ataxin-1. Purkinje cells overexpressing this protein developped a similar pathological process as the one affecting the original SCA1 mouse model, but did not present nuclear inclusions[23,24]. Moreover, a study on primary neuronal cells transfected with various types of HD constructs could not correlate the rate of cell death with the proportion of NIs containing cells[25]. Finally, in

Drosophila models, when the polyglutamine-mediated neurodegeneration was suppressed by the overexpression of protein chaperones from the Hsp40 or 70 families, rescue occured while NIs were still present[15,16].

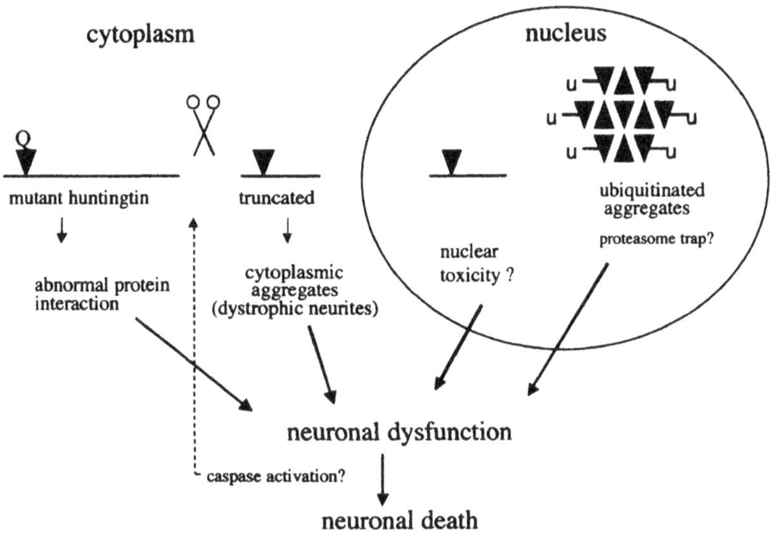

Figure 1. Aggregation of huntingtin in neuronal cytoplasm and nucleus.

Current observations strongly support that polyglutamines induce cellular dysfunctions before cell death. This is difficult to study in man as most samples studied are from patients at the end of many years of disease progression. The mouse models are thus invaluable. In SCA1 mice, the ataxia phenotype and decreased arborisation of Purkinje cells occur much before loss of Purkinje cells[26]. In the R6 HD model, although the neurological symptomatology is very severe, there are only discrete signs of neuronal death, and no widespread astrogliosis. The 20 % brain weight loss observed in these mice is currently unexplained[11,27]. Moreover, we do not observe massive cell loss in our SCA7 mouse model (see below) that however displays very severe functional and histological alterations.

In order to analyse the dysfunctional state of affected neurons, expression profiling analysis or use of substracted cDNA libraries have been performed[24,28]. Downregulation of numerous genes was observed at early stages of pathogenesis. This suggests that transcriptional alterations may be part of the pathological mechanisms, which may be correlated to interaction between mutated huntingtin and the transcriptional factor CBP[29]. It has also been proposed that activation of some caspases may contribute to neuronal dysfunction[30,31].

5. THE MYSTERY OF THE SPECIFICITY OF NEURODEGENERATION

The ubiquitous pattern of expression of the target proteins in the various diseases contrasts with the regional specificity of neurodegeneration. For instance, huntingtin is expressed at high levels in Purkinje cells that are not affected in adult onset HD. However the same cells are the most sensitive to the SCA1 mutation. It was then proposed that abnormal interactions between the mutated protein and some neuron specific proteins could account for the selective pathology. Using a genetic strategy (the yeast two-hybrid system) to screen for protein interacting with ataxin 1, Matilla et al. have found a leucine rich protein (LANP) that interacts much better with an ataxin 1 with 82 repeats than with a normal protein with 30 gln[32]. This protein is expressed predominantly in Purkinje cells, the primary pathological target in SCA1, and colocalizes with ataxin 1 in the nuclear matrix. The authors propose that the presence of mutated ataxin may interfere with the normal function of the LANP protein[32]. Another indication that nuclear matrix may be affected is provided by the abnormal distribution of the PML protein (a nuclear matrix protein associated with nuclear bodies) in Cos cells overexpressing mutated ataxin[14]. It remains to be seen whether other proteins showing sensitivity to polyglutamine length and appropriate neuronal specificity of expression can be found to interact with mutated proteins in other diseases. Proteins whose interaction with huntingtin are modulated by the number of glutamines have been described (HAP1, HIP1, the transcription factor CBP and others), but they do not show a specificity of expression in the neurons selectively affected in HD[29,33,34]. For a review on proposed interactors of huntingtin, see Jones[35]. It should be noted however, that NIs and a neurological phenotype, without obvious neurodegeneration, were obtained by inserting a long CAG/polyglutamine repeat within the mouse HPRT (Hypoxanthin Phosphoribosyl Transferase) gene[36]. This gene is not associated to a polyglutamine expansion disease in man. This suggests that the existence of specific interactors may not be mandatory to elicit polyglutamine mediated neuronal dysfunction.

Alternate (not necessarily exclusive) explanations for the regional selectivity could lie in subtle differences in expression of the target proteins in different neuronal populations, or in differences in the processing of these proteins, that would produce truncated fragments with a higher aggregation potential or toxicity in some neurons. An argument in favor of the first hypothesis is provided by the similarity in the regions affected in HD patients and in the R6 HD mouse model that expresses a highly truncated protein under the control of the promoter region of the human HD gene[12]. It was also reported that within the striatum, the selective vulnerability of different neuron types appears correlated with the normal level of huntingtin expression in the corresponding rat neurons[37]. It is well known that the kinetics of protein aggregation can be highly influenced by rather small differences in protein concentration[10,38]. It has been suggested on the other hand that the expanded polyglutamine may lead to increased sensitivity of huntingtin to cleavage by apoptosis related proteases such as caspase 3[30]. However, the cleavage site accounting for the truncated fragments that accumulates in NIs is upstream of the caspase 3 site of huntingtin[22] (see below).

6. AN INDUCIBLE CELLULAR MODEL FOR ANALYSIS OF PROTEOLYTIC PROCESSING OF HUNTINGTIN

We have devised a cellular model where expression of full length or truncated huntingtin, with normal or expanded polygln repeats, is under the control of a tetracycline inducible promoter (through binding of tetracycline or its analogue doxycycline to a bacterial transactivator)[22]. The cell line we have chosen is a neuroblastoma/glioma hybrid line (NG108) that expresses a neuron-like phenotype upon differentiation. These cells can be studied at least 16 days after induction of huntingtin expression and differentiation, ie much longer than in transient transfection systems. For a review on cellular models, see[22].

Using this model, we could show the progressive formation of nuclear and cytoplasmic aggregates, depending on the length of the recombinant protein (full-length or truncated) and of the polygln tract. In this system, NIs were efficiently formed after 12 days from full-length huntingtin carrying 116 glutamines, while very few NIs but more cytoplasmic inclusions (CIs) were formed with the same construct carrying only 73 repeats. The polyglutamine-containing CIs and NIs contain however only an N-terminal fragment of huntingtin, detected by antibodies against epitopes flanking on either side the polyglutamine repeat. A monoclonal antibody (4C8) recognizing an epitope proximal and upstream of the caspase-3 cleavage site does not detect the inclusions in these cells, nor in patients brain. This indicates that a cleavage occurs at a site proximal to the caspase-3 site, both in the cell model and in patients. We are using this system to further locate the precise cleavage site and identify the responsible protease, as it may provide a potential therapeutic target.

7. MODELLING POLYGLUTAMINE TOXICITY IN THE MOUSE RETINA

Among the polyglutamine disorders, SCA7 is the only one to display degeneration in both brain and retina. Furthermore it shows the greatest sensitivity to polyglutamine length (apart from SCA6 where a different toxic mechanism may be involved, see above). This appears favorable for generating detectable pathology over the life span of the mouse, using a full length construct. We have thus generated transgenic mice overexpressing full-length human ataxin-7 (harbouring 10 or 90 glutamines) in rod photoreceptors, by using the rhodopsin promoter. In this model, expression of the mutant protein causes vision deficiencies and early onset degenerative changes in the retina.

We have raised antibodies against both extremities of human ataxin-7 and showed that an N-terminal immunoreactivity progressively accumulates in the nucleus of photoreceptors leading to the formation of nuclear inclusions (NIs). These NIs are ubiquitinated and recruit a distinct set of chaperone/proteasome subunits. NIs are not immunoreactive for C-terminal ataxin-7 antibodies, suggesting a proteolytic processing of ataxin-7[Q90]. These molecular features are not attributable to the particularities of the photoreceptor cell type : we could confirm them in another SCA7 mouse model which expresses the protein in Purkinje cells. Such a proteolytic step may therefore be involved in SCA7 as well as in HD and dentatorubral-pallidoluysian atrophy[18,22,40]. Overexpression of ataxin-7[Q90] in rods caused a severe retinal degeneration, including loss of outer segments and disorganization of the outer nuclear layer. Pathological changes appear at about 1 month of age and are correlated to an abnormal electroretinogramm response *in vivo*. Alterations are not limited to the targeted cell type: horizontal and rod-bipolar cells, the post-synaptic partners of rods, develop morphological changes although they do not express the mutation. These observations are the first clear evidence that trans-neuronal responses are part of polyglutamine toxicity. Because retina is a suitable tissue for gene transfer and drug delivery, our model offers new opportunities to test therapeutic strategies for triplet repeat disorders.

8. NEW HOPES FOR THERAPY

Recent discoveries have raised exciting new hopes for treatment that may delay or suppress the polyglutamine-mediated toxicity. Several neurotrophic factors and anti-apoptotic agents have shown protective effects on cellular models. For a review see[39]. The aggregation properties of mutated proteins provide also a potential target and efforts have been directed towards identification of small molecules that could decrease fibrillary formation in vitro[41]. The availability of animal models allows *in vivo* assessment of therapeutic approaches. One study reported that the neurological phenotype of R6 HD mice can be delayed by crossing them with transgenic animals expressing a dominant negative form of caspase-1, or by treatment with a caspase inhibitor[31]. Further hopes arised from fly geneticists. Two groups have reported that when flies express either hsp70 or hsp40 chaperones together with a polyglutamine-coding transgene, the severe neurodegenerative phenotype of the model is suppressed[15,42]. However, treatments may encounter enormous difficulties if the disease progression is irreversible after its onset. A recent report based on an inducible mouse model, showed that even after the appearance of symptoms, if the expression of the HD mutation is turned off, the animals recover[43]. This observation suggests that the pathogenic process is not irreversibly triggered by the mutation, but is rather continuous and therefore is more likely to be blocked by effective treatments.

ACKNOWLEDGEMENTS

Studies in our laboratory are supported by CNRS, INSERM, CHRU Strasbourg and Fondation Louis Jeantet. A.L and G.Y. are supported by the Hereditary Disease Foundation and the Fondation pour la Recherche Medicale, respectively.

REFERENCES

1. David G, Abbas N, Stevanin G et al. (1997) Cloning of the SCA7 gene reveals a highly unstable CAG repeat expansion. Nat Genet 17: 65-70.
2. Koide R, Kobayashi S, Shimohata T et al. (1999) A neurological disease caused by an expanded CAG trinucleotide repeat in the TATA-binding protein gene: a new polyglutamine disease? Hum Mol Genet 8: 2047-2054.
3. The Huntington's Disease Collaborative Research Group (1993) A novel gene containing a trinucleotide repeat that is expanded and unstable on Huntington's disease chromosomes. Cell 72: 971-983.
4. Gusella JF, McNeil S, Persichetti F et al. (1996) Huntington's disease: Cold Spring Harbor Symposia on Quantitative Biology.
5. Zoghbi HY (1997) CAG repeats in SCA6: anticipating new clues. Neurology 49: 1-5.
6. Trottier Y, Lutz Y, Stevanin G et al. (1995) Polyglutamine expansion as a pathological epitope in Huntington's disease and four dominant cerebellar ataxias. Nature 378: 403-406.

7. Lunkes A, Trottier Y, Fagart J et al. (1999) Properties of polyglutamine expansion in vitro and in a cellular model for Huntington's disease. Philos Trans R Soc Lond [Biol] 354: 1013-1019.
8. Scherzinger E, Lurz R, Turmaine M et al. (1997) Huntingtin-encoded polyglutamine expansions form amyloid-like protein aggregates in vitro and in vivo. Cell 90: 549-558.
9. Stott K, Blackburn JM, Butler PJ et al. (1995) Incorporation of glutamine repeats makes protein oligomerize: implications for neurodegenerative diseases. Proc Natl Acad Sci USA 92: 6509-6513.
10. Scherzinger E, Sittler A, Schweiger K et al. (1999) Self-assembly of polyglutamine-containing huntingtin fragments into amyloid-like fibrils: Implications for Huntington's disease pathology. Proc Natl Acad Sci USA 96: 4604-4609.
11. Mangiarini L, Sathasivam K, Seller M et al. (1996) Exon 1 of the *HD* gene with an expanded CAG repeat is sufficient to cause a progressive neurological phenotype in transgenic mice. Cell 87: 493-506.
12. Davies SW, Turmaine M, Cozens BA et al. (1997) Formation of neuronal intranuclear inclusions underlies the neurological dysfunction in mice transgenic for the HD mutation. Cell 90: 537-548.
13. Burright EN, Clark HB, Servadio A et al. (1995) SCA1 transgenic mice: a model for neurodegeneration caused by an expanded CAG trinucleotide repeat. Cell 82: 937-948.
14. Skinner PJ, Koshy B, Cummings CJ et al. (1997) SCA1 pathogenesis involves alterations in nuclear matrix associated structures. Nature 389: 971-974.
15. Kazemi-Esfarjani P, Benzer S (2000) Genetic suppression of polyglutamine toxicity in Drosophila. Science 287: 1837-1840.
16. Warrick JM, Paulson HL, Gray-Board GL et al. (1998) Expanded polyglutamine protein forms nuclear inclusions and causes neural degeneration in Drosophila. Cell 93: 939-949.
17. Becher MW, Kotzuk JA, Sharp AH et al. (1998) Intranuclear neuronal inclusions in Huntington's Disease and Dentatorubral and pallidoluysian atrophy: correlation between the density of inclusions and *IT15* CAG triplet repeat length. Neurobiol Dis 4: 387-397.
18. Di Figlia M, Sapp E, Chase KO et al. (1997) Aggregation of Huntingtin in neuronal intranuclear inclusions and dystrophic neurites in brain. Science 277: 1990-1993.
19. Koyano S, Uchihara T, Fujigasaki H et al. (2000) Neuronal intranuclear inclusions in spinocerebellar ataxia type 2. Ann Neurol 47: 550.
20. Paulson HL, Perez MK, Trottier Y et al. (1997) Intranuclear inclusions of expanded polyglutamine protein in spinocerebellar ataxia type 3. Neuron 19: 333-344.
21. Gutekunst CA, Li SH, Yi H et al. (1999) Nuclear and neuropil aggregates in Huntington's disease: relationship to neuropathology. J Neurosci 19: 2522-34.
22. Lunkes A, Mandel JL (1998) A cellular model that recapitulates major pathogenic steps of Huntington's disease. Hum.Mol.Genet. 7: 1355-1361.
23. Klement IA, Skinner PJ, Kaytor MD et al. (1998) Ataxin-1 nuclear localization and aggregation: role in polyglutamine-induced disease in SCA1 transgenic mice. Cell 95: 41-53.
24. Lin X, Antalffy B, Kang D et al. (2000) Polyglutamine expansion down-regulates specific neuronal genes before pathologic changes in SCA1. Nat Neurosci 3, 157-63.
25. Saudou F, Finkbeiner S, Devys D et al. (1998) Huntingtin acts in the nucleus to induce apoptosis but death does not correlate with the formation of intranuclear inclusions. Cell 95: 55-66.

26. Clark HB, Burright EN, Yunis WS et al. (1997) Purkinje cell expression of a mutant allele of SCA1 in transgenic mice leads to disparate effects on motor behaviors followed by a progressive cerebellar dysfunction and histological alterations. J Neurosci 17: 7385-7395.
27. Turmaine M, Raza A, Mahal A et al. (2000) Nonapoptotic neurodegeneration in a transgenic mouse model of Huntington's disease. Proc Natl Acad Sci USA 97: 8093-8097.
28. Luthi-Carter R, Strand A, Peters NL et al. (2000) Decreased expression of striatal signaling genes in a mouse model of Huntington's disease. Hum Mol Genet 9: 1259-1271.
29. Steffan JS, Kazantsev A, Spasic-Boskovic O et al. (2000) The Huntington's disease protein interacts with p53 and CREB-binding protein and represses transcription. Proc Natl Acad Sci USA 97: 6763-6768.
30. Goldberg YP, Nicholson DW, Rasper DM et al. (1996) Cleavage of huntingtin by apopain, a proapoptotic cysteine protease, is modulated by the polyglutamine tract. Nat Genet 13: 442-449.
31. Ona VO, Li M, Vonsattel JP et al. (1999) Inhibition of caspase-1 slows disease progression in a mouse model of Huntington's disease. Nature 399: 263-267.
32. Matilla A, Koshy B, Cummings CJ et al. (1997) The cerebellar leucine rich acidic nuclear protein (LANP) interacts with ataxin-1. Nature 389: 974-978.
33. Li XJ, Sharp AH, Li SH et al. (1996) Huntingtin associated protein (HAP1): Discrete neuronal localizations in the brain resemble those of neuronal nitric oxide synthase. Proc Natl Acad Sci USA 93: 4839-4844.
34. Wanker EE, Rovira C, Scherzinger E et al. (1997) HIP-I: a huntingtin interacting protein isolated by the yeast two- hybrid system. Hum Mol Genet 6: 487-95.
35. Jones L (2000). Huntingtin-interacting Proteins and their Relevance to Huntington's Disease Etiology. Neurosci News: 55-63.
36. Ordway JM, Tallaksen-Greene S, Gutekunst CA et al. (1997) Ectopically expressed CAG repeats cause intranuclear inclusions and a progressive late onset neurological phenotype in the mouse. Cell 91: 753-763.
37. Kosinski CM, Cha JH, Young AB et al. (1997) Huntingtin immunoreactivity in the rat neostriatum: differential accumulation in projection and interneurons. Exp.Neurol 144: 239-247.
38. Jarrett JT, Lansbury PTJ (1993) Seeding "one-dimensional cristallization" of amyloid: a pathogenic mechanism in Alzheimer's disease and Scrapie ? Cell 73: 1055-1058.
39. Lunkes A, Mandel JL (2000) Cellular Models of Huntington's Disease. Neurosci News: 30-37.
41. Heiser V, Scherzinger E, Boeddrich A et al. (2000) Inhibition of huntingtin fibrillogenesis by specific antibodies and small molecules: implications for Huntington's disease therapy. Proc Natl Acad Sci USA 97: 6739-44.
40. Schilling G, Wood JD, Duan K et al. (1999) Nuclear accumulation of truncated atrophin-1 fragments in a transgenic mouse model of DRPLA. Neuron 24: 275-86.
42. Warrick JM, Chan HYE, Gray-Board GL et al. (1999) Suppression of polyglutamine-mediated neurodegeneration in Drosophila by the molecular chaperone Hsp70. Nat Genet 23: 425-428.
43. Yamamoto A, Lucas JJ, Hen R (2000) Reversal of neuropathology and motor dysfunction in a conditional model of Huntington's disease. Cell 101: 57-66.

DEMENTIA IN THE NEURONAL CEROID-LIPOFUSCINOSES

[1]Hans H. Goebel and [2]Alfried Kohlschütter
[1]*Department of Neuropathology, Johannes Gutenberg University, Mainz, Germany,*
[2]*Department of Pediatrics, University of Hamburg, Hamburg, Germany*

1. INTRODUCTION

Dementia is defined as a decline in cognitive abilities such as impairment of memory, reasoning, behaviour, attention, motivation and effectiveness. The term usually implies that normal mature mental capability was achieved before, and it is therefore mostly ascribed to adult patients.

In the pediatric age group, however, dementia also occurs in a variety of progressive neurological disorders. It has to be differentiated from mental retardation, a term that implies mental development proceeding at a slower than normal pace but without loss of mental capabilities already acquired. Early in the course of a neurodegenerative disorder, an unexpected delay of mental maturation may, to the inexperienced, suggest mental retardation. At a later stage, mental regression (dementia) will become evident. Dementia may be difficult to recognize when the course of the disease is complicated by other prominent symptoms such as seizures, visual impairment, and motor abnormalities.

Among the progressive and often fatal neurodegenerative diseases in children, the neuronal ceroid-lipofuscinoses (NCL) form the largest nosological group. An early classification subdivided the NCL group according to age at clinical onset into infantile, late-infantile, early-juvenile, juvenile and adult types[1], while a new gene-based classification so far distinguishes eight different forms, CLN1 to CLN8 (Table 1)[2]. The old clinically oriented classification, however, is still of practical usefulness in the first approach to diagnosing a disorder associated with dementia in the pediatric age group.

Type	Chromosome	Gene product	Clinical type
CLN1	1p32	lysosomal palmitoyl protein thioesterase:	INCL
CLN2	11p15	lysosomal tripeptidyl peptidase I:	LINCL
CLN3	16p12	lysosomal transmembraneous CLN3 protein (battenin):	JNCL
CLN4	not known	*not known*	ANCL
CLN5	13q31-32	transmembraneous CLN5 protein:	LINCL, Finnish variant
CLN6	15q21-23	*not known*	EJNCL, Indian variant
CLN7	not known	*not known*	LINCL, Turkish variant
CLN8	8	transmembraneous CLN8 protein:	"Northern" epilepsy

Table 1. Current genetic status of NCL

2. DEMENTIA IN NCL

Infantile NCL. — In patients with this form of NCL, mental deterioration during the second year of life is particularly striking, provided one does not overlook the previous normal early development of the child. Rapid loss of all mental and motor capabilities ensues, eventually leading to a long-lasting vegetative state[3].

Late-infantile NCL. — Mental regression in this form occurs around 2-4 years of life with loss of language, behavioural abnormalities and general intellectual decline[4].

Juvenile NCL. — In this form, dementia follows the loss of visual acuity. It becomes fully recognizable during the second decade of life and progresses slowly. Actually, the decline in intellect represents an important parameter to monitor progression of JNCL and to identify patients afflicted with the protracted variant[5]. It is associated with a peculiar psycho-organic syndrome characterized by feelings of frustration, depression and aggressive behaviour.

Northern epilepsy – CLN8. — This newly recognized type of NCL also commences in childhood and shows dementia from the second decade onwards[6]. It progresses very slowly and patients live well into adulthood thus representing a very protracted form of childhood NCL[7].

Adult NCL. — In patients with this form, loss of mental abilities starts at around 30 years of age and is similar to more well-known other dementias[8]. Two clinical phenotypes, A and B (Table 2), have been separated[9], dementia being a component of type B.

Type A	Type B
▪ Myoclonic Seizures	▪ Dementia
▪ Neuropsychiatric changes	▪ Motor abnormalities
▪ Ataxia	▪ TIC-like facial dyskinesia
▪ Dysarthria	
Age of onset: 29.0 ± 10.5 yr	Age of onset: 12-47 yr, mean 30.9 ± 9.8 yr
Duration of illness: 12.4 ± 9.4 yr	Duration of illnes: 7-25 yr, mean 12.6 ± 5.5 yr

Table 2. Clinical types of adult NCL[9]

3. MORPHOLOGY OF NCL

Morphologically, the NCL are marked by loss of neurons and intracellular accumulation of lipopigments. The loss of neurons, both in the cerebrum and the cerebellum, seems to parallel onset and velocity in progression of the clinical subtype of the disease.

In infantile NCL the flat EEG indicates severe cerebrocortical damage which, at autopsy, may reveal almost complete neuronal depletion of cerebral and cerebellar cortices (Figure 1). A similar neuronal devastation may occur in late-infantile NCL, but Bielschowsky's original patients still had a considerable population of cerebrocortical neurons left (Figure 2)[10-12].

In juvenile NCL, neuronal loss in the cerebral cortex is less pronounced and, based on the pigmentoarchitectonic technique, appears to be selective in that small stellate neurons of layer II and pyramidal neurons of layer V are preferentially reduced in number[13-15].

In adult NCL, neuronal loss is even less advanced than in juvenile NCL and it has been described as diffuse[16], or selective following the pattern of that seen in juvenile NCL[2,17-18]. Loss of neuronal pigment-laden stellate cells in layers II and III, but only mild neuronal depletion in layer Vb, were found.

4. COMMENT

Clinicopathological correlations concerning dementia have not yet been performed in NCL, because quantitation of dementia as well as of morphological lesions is lacking.

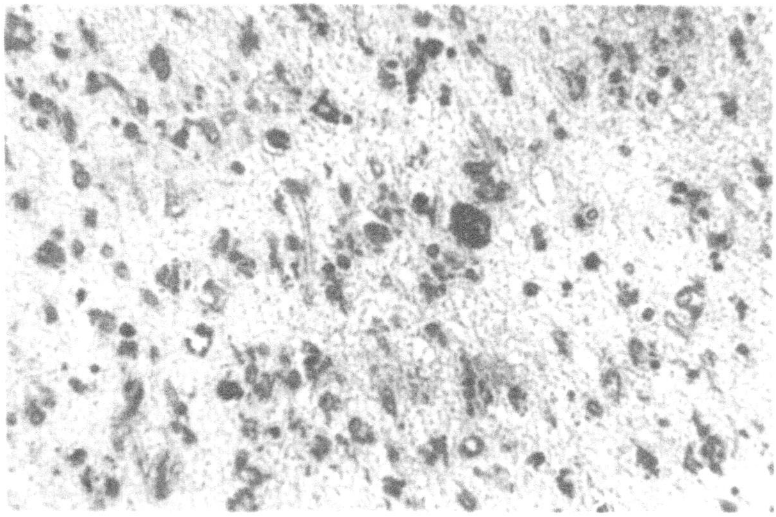

Figure 1. Severe neuronal depletion and glial proliferation in the cerebral cortex of infantile NCL, PAS-hematoxylin, X 330.

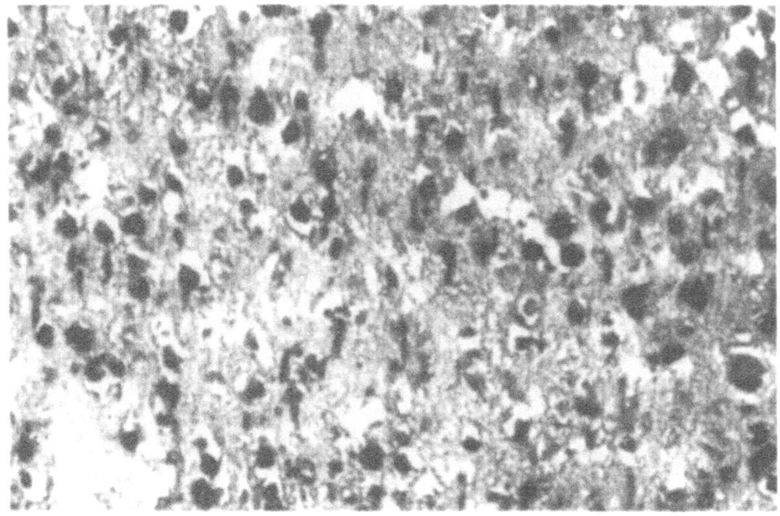

Figure 2. Preservation of a large number of neurons marked by accumulation of lipopigments containing subunit c of mitochondrial ATP synthase (antibody by courtesy of Prof. Kominami/Tokyo), cerebral cortex in late-infantile NCL originally described by M. Bielschowsky[10-11], immunohistochemical reaction, X 140.

Whether a particular class of neurons or a particular brain region is affected by cortical neuronal depletion remains to be determined. Likewise, it is not clear whether the accumulation of lipopigments causes neurodegeneration and dementia. An alternative explanation for the neurodegenerative process is one of mitochondrial disturbance that affects specific neurons with high mitochondrial energy metabolism[19].

In infantile NCL, it is safe to assume that near-complete neuronal depletion of the cerebral cortex parallels profound dementia, but what morphologically parallels the clinical onset of dementia in this form of NCL is unknown.

In juvenile NCL, dementia is profound in the end-stage but, morphologically, lipopigment formation in cerebrocortical neurons is very conspicuous whereas loss of cerebrocortical neurons is moderate and far less advanced than in infantile or late-infantile NCL. Thus, lipopigment formation within cerebrocortical neurons may, indeed, contribute to dementia. Hence, quantifying studies of the neuronal populations in the cerebral cortex and subcortical grey matter are necessary.

CLN1:	GROD
CLN2:	curvilinear
CLN3:	vacuolar fingerprint
CLN4:	*not known*
CLN5:	non-vacuolar fingerprint and granular
CLN6:	non-vacuolar fingerprint and granular
CLN7:	non-vacuolar fingerprint and granular
CLN8:	*not known*

Table 3: Genotype –(morpho)Phenotype Correlations (lymphocytes)

Precise nosological classification of dementia in the adult can only reliably be ascertained by molecular genetic investigations in familial forms of dementia, by biochemical studies in neurometabolic types of dementia, in sporadic neurodegenerative conditions of dementia by autopsies. Conversely, certain neurometabolic and neurodegenerative dementing disorders, among them the neuronal ceroid-lipofuscinoses in children, may be recognized during life by extracerebral biopsy and subsequent electron microscopic studies of such tissues as skin, conjunctiva, rectum, skeletal muscle or circulating lymphocytes. Of these diverse tissues, lymphocytes provide the best genotype-phenotype correlation among the childhood forms of NCL (Table 3). While lymphocytes have not been reported to be affected in adult NCL, accruing lipopigments encompass the entire known spectrum of ultrastructural patterns (Table 4) in adult NCL.

▪ *Granular:*	neurons
▪ *Fingerprint:*	neurons, eccrine sweat glands, smooth/muscle cells
▪ *Recti-/curvilinear:*	muscle fibres, splenic cells

Table 4. Morphologic spectrum of adult NCL/CLN4

ACKNOWLEDGEMENT

Support by the European Concerted Action ECA-"NCL", project No. BMH 4-CT 95-0563, BIOMED 2 programme, photography by W. Meffert and editorial assistance by Mrs. A. Wöber are gratefully acknowledged.

REFERENCES

1. Zeman W, Donahue S, Dyken P. et al. (1970) The neuronal ceroid-lipofuscinoses (Batten-Vogt syndrome). In: Handbook of Clinical Neurology (Vinken PJ, Bruyn GW eds), pp 588-679, Vol 10: North-Holland Publishing Company, Amsterdam.
2. Goebel HH, Mole SE, Lake BD (eds.) (1999) The Neuronal Ceroid Lipofuscinoses (Batten Disease). IOS Press, Amsterdam.
3. Santavuori P, Gottlob I, Haltia M et al. (1999) CLN1 - Infantile and other types of NCL with GROD. In: The Neuronal Ceroid Lipofuscinoses. (Batten Disease) (Goebel HH, Mole SE, Lake BD eds), pp 16-36: IOS Press, Amsterdam.
4. Williams RE, Gottlob I, Lake BD et al. (1999) CLN2 - Classic late infantile NCL. In: The Neuronal Ceroid Lipofuscinoses (Batten Disease) (Goebel HH, Mole SE, Lake BD eds), pp 37-54: IOS Press, Amsterdam.
5. Kohlschütter A, Laabs R, Albani M (1988) Juvenile neuronal ceroid lipofuscinosis (JNCL): quantitative description of its clinical variability. Acta Paediatr Scand 77: 867-872.
6. Haltia M, Tyynelä J, Hirvasniemi A et al. (1999) CLN 8 - Northern epilepsy. In: The Neuronal Ceroid Lipofuscinoses (Batten Disease) (Goebel HH, Mole SE, Lake BD eds), pp 117-124: IOS Press, Amsterdam.
7. Herva H, Tyynelä J, Hirvasniemi A et al. (2000) Northern epilepsy: a novel form of neuronal ceroid-lipofuscinosis. Brain Pathol 10: 215-222.
8. Martin JJ, Gottlob I, Goebel HH et al. (1999) CLN4 - Adult NCL. In: The Neuronal Ceroid Lipofuscinoses (Batten Disease) (Goebel HH, Mole SE, Lake BD eds), pp 77-90: IOS Press, Amsterdam.
9. Berkovic SF, Carpenter S, Andermann F et al. (1988) Kufs' disease: a critical reappraisal. Brain 111: 27-62.
10. Bielschowsky M (1913) Über spätinfantile familiäre amaurotische Idiotie mit Kleinhirnsymptomen. Dtsch Zeitschr Nervenheilk 50: 7-29.
11. Bielschowsky M (1920) Zur Histopathologie und Pathogenese der amaurotischen Idiotie mit besonderer Berücksichtigung der zerebellaren Veränderungen. Journal für Psychologie und Neurologie 26: 129-199.
12. Goebel HH, Gerhard L, Kominami E et al. (1996) Neuronal ceroid-lipofuscinosis – late-infantile or Jansky-Bielschosky type – revisited. Brain Pathol. 6: 225-228.
13. Braak H, Goebel HH (1978) Loss of pigment-laden stellate cells: a severe alteration of the isocortex in juvenile neuronal ceroid-lipofuscinosis. Acta Neuropathol 42: 53-57.
14. Braak H, Goebel HH (1979) Pigmentoarchitectonic pathology of the isocortex in juvenile neuronal ceroid lipofuscinosis: Axonal enlargements in layer IIIab and cell loss in layer V. Acta Neuropathol 46: 79-83.
15. Braak H, Braak E (1993) Pathoarchitectonic pattern of iso- and allocortical lesions in juvenile and adult neuronal ceroid-lipofuscinosis. J Inher Metab Dis 16: 259-262.

16. Iseki E, Amano N, Yokoi S et al. (1987) A case of adult neuronal ceroid-lipofuscinosis with the appearance of membranous cytoplasmic bodies localized in the spinal anterior horn. Acta Neuropathol 72: 362-368.
17. Goebel HH, Braak H (1989) Adult neuronal ceroid-lipofuscinosis. Clin Neuropathol 8: 109-119.
18. Martin JJ, Libert J, Ceuterick C (1987) Ultrastructure of brain and retina in Kufs' disease (adult type-ceroid-lipofuscinosis). Clin Neuropathol 6: 231-235.
19. Jolly RD, Kohlschütter A, Palmer DN et al. (1999) The Neuronal Ceroid-Lipofuscinoses (Batten Disease). In: Advances in Cell Aging and Gerontology. Genetic Aberrancies and Neurodegenerative Disorders (Mattson MP ed), pp 391-420, Vol 3: JAI Press, Stamford, Connecticut.

NEURODEGENERATION-ASSOCIATED PROTEINS AND INFLAMMATION IN SPORADIC INCLUSION-BODY MYOSITIS

[1]Johannes B. Lampe, [2]Maggie C. Walter and [1]Heinz Reichmann
[1]Department of Neurology, Technical University of Dresden, Dresden, Germany, [2]Friedrich Baur Institute Munich, Munich, Germany

1. INTRODUCTION

The same proteins accumulate in the brains of patients with sporadic or hereditary Alzheimer's disease and in the muscle fibres of patients with sporadic inclusion-body myositis (s-IBM) and hereditary inclusion-body myopathy (h-IBM). The pathogenic mechanisms underlying both disorders may be similar. Therefore, IBM may serve as an easily accessible peripheral model for the aetiology of Alzheimer's disease[1].

S-IBM is diagnosed in patients with muscle weakness affecting both proximal and distal muscles of arms and legs. Clinical characteristics are early involvement of distal muscles, in particular finger flexors and foot extensors, and asymmetric muscle atrophy predominantly in quadriceps muscles and arm flexors[2]. Frequently, patients additionally suffer from mild dysphagia[3]. Typically, s-IBM patients are older than 50 years of age. However, young adults or children may be affected too[4]. In addition, there is a 3:1 male preponderance[5,6].

Histologically, s-IBM is characterised by an inflammatory myopathy with predominantly CD8 positive monocytic infiltrates accompanied by vacuolated muscle fibres and intracellular amyloid deposits (Figure 1). Electron microscopy reveals deposits of 15-18 nm tubulofilaments and

Neuropathology and Genetics of Dementia, Edited by Tolnay and Probst
Kluwer Academic/Plenum Publishers, New York 2001

confirms the diagnosis (Figure 2). As a rule, s-IBM is associated with systemic immune disorders or non-specific autoantibodies[7].

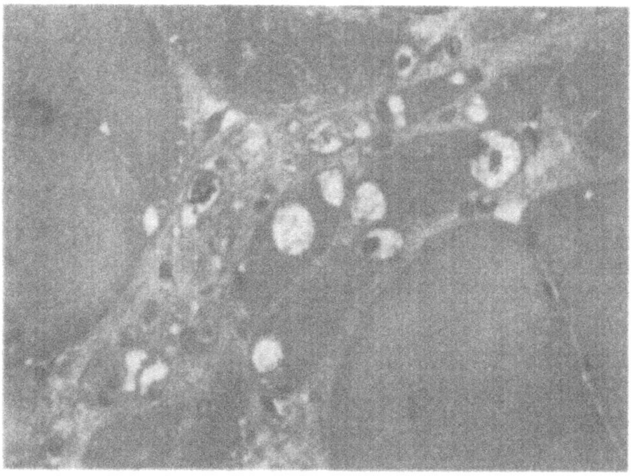

Figure 1: Sporadic inclusion-body myositis: muscle biopsy reveals rimmed vacuoles within muscle fibres. H&E staining, x400.

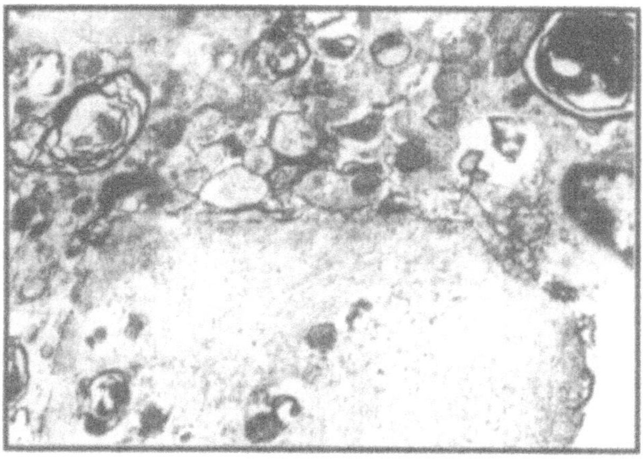

Figure 2: Electronmicroscopically rimmed vacuoles are autophagic vacuoles with myelin-configured catabolic products. Additionally, filamentous bundled structures prove the diagnosis of inclusion body myositis (by courtesy of Prof. J. Müller-Höcker, Institute of Pathology, University of Munich)

1.1 Neurodegeneration-associated proteins and s-IBM

Several proteins are found in the inclusions in s-IBM including epitopes of beta-amyloid precursor protein (βAPP), phosphorylated tau, alpha-1-

antichymotrypsin, apolipoprotein E, prion protein and presenilin-1[8-10]. Interestingly, these proteins are typically found in the brains of patients suffering from different neurodegenerative diseases, especially Alzheimer's disease. The genetic predisposition to most cases of Alzheimer's disease is probably multifactorial. Several studies demonstrated that the e4 allele of apolipoprotein E (Apo E) is associated with familial and sporadic late-onset AD[11]. On the other hand, a protective effect of the ε2 allele has been postulated[12]. Since the typical IBM inclusions contain apolipoprotein E[9] it was suggested that ApoE may be involved in the pathogenesis of IBM. In fact, one group found that the ApoE ε4 may not only predispose to Alzheimer's disease but also to the development of IBM[13]. However, we have analysed the ApoE allele frequency in 23 s-IBM patients and found the ApoE ε4 to be only slightly increased in comparison with normal controls (0.18 vs. 0.11)[14] (Lampe et al., unpublished data). These findings are in accordance with the results of another group[15]. Nevertheless, higher numbers of patients would have to be analysed to clarify this point.

The zygosity at the polymorphic codon 129 of the prion protein influences the clinical and histopathological appearance of prion diseases in different ways. A subtype of familial Creutzfeldt-Jakob disease (CJD178) and fatal familial insomnia (FFI) are linked to the same mutation at prion gene codon 178 (Asn178). However, both diseases display distinct clinical and histological phenotypes. The genotypic basis for the difference between FFI and CJD178 depends on the zygosity at codon 129 of the mutant prion protein allele: 129Val plus 178Asn leads to CJD, whereas 129Met plus 178Asn results in FFI[16]. Moreover, homozygosity at prion gene codon 129 predisposes to the development of sporadic and iatrogenic Creutzfeldt-Jakob-disease[17-21]. Additionally, all patients described so far with the new variant of CJD have been homozygous for methionine[22].

Analysing the polymorphic prion gene codon 129 in 31 s-IBM patients, we found methionine homozygosity in more than 61% of the cases compared to 40% in normal control. Using metaanalysis with data from three different western European countries, this was statistically significant[23,24]. In one s-IBM patient out of 31 we identified a new prion gene mutation at codon 140, which led to histidine being substituted by arginine. The neuropsychological examination of this patient was normal and there is no family history of neuropsychiatric diseases. Relatives, three maternal and one paternal, did not display this novel prion mutation. However, a causal link between this mutation and s-IBM in that patient has not yet been proven (Lampe et al., unpublished data).

1.2 Is Sporadic Inclusion-Body Myositis a Prion Disease of Skeletal Muscle?

Several features may link the prion gene and prion-gene products to skeletal muscle. Therefore, it has been speculated that IBM represents a human prion disease of the skeletal muscle[25,26]: (i) One patient with Gerstmann-Sträussler-Scheinker-syndrome[27,28] was described as having vacuolation and ragged-red fibres in the muscle fibres. (ii) Several authors reported myopathic changes resembling polymyositis in sheep suffering from scrapie[29-31]. (iii) DeArmond and Prusiner described elevated levels of proteinase K-resistant PrP in IBM in two review articles[26,32]. This would be of special interest, since the detection of proteinase K-resistant PrP is diagnostic for prion diseases of the central nervous system[33]. However, the authors have never published any original data on this.

There are different arguments against a similar aetiology of s-IBM and spongiform encephalopathies. (i) Several neurodegeneration-related proteins are found in s-IBM. In contrast, prion protein is predominantly detectable in spongiform encephalopathies[8-10]. (ii) In contrast to prion diseases, prion gene expression is upregulated in IBM[34]. (iii) To our best knowledge there are no reports on inoculation experiments in IBM in the literature. However, this would be very important for excluding the infectivity of skeletal muscle in IBM. To summarise, the better arguments speak out against a prion disease aetiology of s-IBM.

However, in our opinion aetiology and pathogenesis of s-IBM is linked to the prion gene: (i) Prion protein accumulates and prion gene mRNA is upregulated in s-IBM[10,34,35]. (ii) Moreover, transgenic mice overexpressing PRNP suffer from a necrotic myopathy at an older age[36]. (iii) Our own data indicates that PRNP codon 129 methionine homozygosity may predispose for the development of s-IBM and we have recently identified a new mutation in one s-IBM patient[23] (Lampe et al., unpublished data). On the other hand, the hereditary form of the IBM has been linked to chromosomes 9 and 17[37,38]. Since h-IBM lacks inflammation in almost all cases the aetiology in s-IBM and h-IBM may differ.

2. INFLAMMATION AND S-IBM

IBM may appear as s-IBM or hereditary h-IBM myopathies. A locus for an autosomal-recessively transmitted form of h-IBM has been linked to chromosome 9p1-q1[37]. Additionally, a locus for autosomal-dominantly transmitted h-IBM was identified on chromosome 17p13.1[38]. Inclusions in both s-IBM and h-IBM contain amyloid deposits. The aetiology of IBM is unknown. Typically, s-IBM is characterised by inflammation, whereas h-IBM lacks inflammatory changes in most cases[39,40]. Therefore, the inherited form has been termed inclusion-body myopathy to emphasise this contrast to sporadic inclusion-body myositis. Additionally, it has been speculated that the aetiology and pathogenesis of h-IBM and s-IBM may differ. Nevertheless, the degree of inflammation may vary significantly among different s-IBM cases and even be absent in some[39].

Among 99 s-IBM cases Koffman and colleagues found diseases with altered immune functioning in 13 patients. Autoantibodies were detected in 44 patients[7].

In both IBM and polymyositis (PM), cell-mediated immune mechanisms seem to be crucial since CD8+ cytotoxic T-cells infiltrate muscle fibres. Muscle fibres in IBM and PM strongly express HLA class I molecules on the cell surface. Therefore, CD8+ cytotoxic T-cells most probably recognise antigenic epitopes bound to HLA class I molecules[41]. The dissection of HLA class I antigens is of particular interest, since the mononuclear infiltrate in s-IBM patients is dominated by CD8+ T-cells[41]. Moreover, it has been shown, that T cells infiltrating muscle fibres in s-IBM display a restricted usage of variable (V) alpha/beta gene families and certain T-cells may be oligoclonally expanded in s-IBM[42]. These observations lead to the speculation that cytotoxic T-cells may recognise an autoantigen on IBM myotubes[44].

Several studies identified HLA class I and II antigens predisposing to the development of s-IBM[42,44-49]. We could confirm these results in a study of 32 s-IBM patients. Moreover, the typing of HLA class I showed an increase in the A*0301-3N allele. This is a predisposition towards s-IBM that has not yet been described[50]. Since most muscle fibres in IBM strongly express HLA class I molecules on the cell surface and T-cells infiltrating s-IBM show a restricted T-cell receptor usage1, the infiltrating CD8+ cytotoxic T-cells may detect antigens bound to distinct HLA class I molecules[41]. Hence, the aetiology of s-IBM may be triggered by as yet unknown immune mechanisms.

Early electron microscopical studies emphasised that the typical filaments seen in IBM resemble viral nucleocapsids. Therefore, it was suggested that IBM is a slow-virus infection of the skeletal muscle[4,52]. Immunostaining of the inclusions for mumps virus antigen suggested a persistent mumps virus infection as the cause of IBM[53]. However, those reports could not be confirmed in further studies which used PCR, *in-situ* hybridisation and immunocytochemical techniques[54,55]. Interestingly, two HIV-I (human immunodeficiency virus type 1) infected men and one woman infected with HTLV-1 (human T-cell leukaemia virus type 1) with clinical and histological features consistent with the diagnosis of s-IBM were published[56]. Moreover, patients with post-polio syndrome and s-IBM have been reported[57,58]. However, these are only case reports. A common viral cause for s-IBM has not been able to be demonstrated yet.

3. MODEL FOR S-IBM AETIOLOGY AND PATHOGENESIS

According to the "nucleation theory" different amyloidogenic components may serve as a nucleating agent for amyloid formation[59,60]. Moreover, interactions between several proteins found in neurodegenerative diseases and molecular chaperones have been suggested[61]. In this respect, the observation of enhanced expression of alphaB-crystallin in IBM fibres is of special interest[62]. Recently, a patient homozygous for the transthyretin Val122Ile allele with both sporadic inclusion-body myositis and cardiac amyloidosis was reported and we identified a patient with a novel mutation in the prion gene[63] (Lampe et al., unpublished data). This gives rise to the speculation that mutations in different neurodegeneration-related genes may serve as nucleating agents in IBM. Interestingly, overexpression of beta-amyloid precursor protein in cultured normal human muscle, resembles several histological aspects of IBM including congophilia, 6-10 nm filaments, and 15-21 nm tubulofilamentous inclusions in the nuclei[64]. Moreover, transgenic mice overexpressing fragments of the amyloid-beta precursor gene resemble different aspects of IBM pathology[65,66]. Interestingly, the overexpression of the prion protein in transgenic mice may lead to a narcotising myopathy similar to IBM[37].

A model for s-IBM should integrate the different findings made by several investigators in IBM including parameters of oxidative stress[67-69], mitochondrial changes[70], and protein accumulation[71]. We suggest a modified model of the hypothetical model published by Askanas and Engel[1]. The authors propose that mutated genes (h-IBM) or other unknown factors, as in the case of s-IBM, lead to βAPP overexpression. This βAPP overexpression induces oxidative stress and the formation of free radicals. As a direct consequence of oxidative stress COX is downregulated and several other factors are upregulated. These include nitrotyrosine, malondialdehyde, nNOS, iNOS, HO-1, SOD1, glutathione Px, and catalase. Simultaneously mitochondria are disturbed leading to mitochondrial abnormalities. Additionally, the oxidative stress may be responsible for the appearance of vacuolated fibres accompanied by congophilia and the accumulation of neurodegeneration-related proteins such as phosphorylated tau, ubiquitin, presenilin 1, α1ACT, ApoE and PrP.

REFERENCES

1. Askanas V, Engel WK (1998) Does overexpression of betaAPP in aging muscle have a pathogenic role and a relevance to Alzheimer's disease? Clues from inclusion body myositis, cultured human muscle, and transgenic mice. Am J Pathol 153: 1673-1677.
2. Dalakas MC (1995) Update on the use of intravenous immune globulin in the treatment of patients with inflammatory muscle disease. J Clin Immunol 15: 70S-75S.
3. Dalakas MC, Sonies BC, Lopez-Devine J et al. (1996) Dysphagia as a frequent manifestation in patients with inclusion body myositis. Neurology 46: Suppl.
4. Chou SM (1993) Inclusion body myositis. Baillières Clin Neurol 2: 557-577.
5. Dalakas MC (1991) Polymyositis, dermatomyositis and inclusion-body myositis. N Engl J Med 325: 1487-1498.
6. Sekul EA, Dalakas MC (1993) Inclusion body myositis: new concepts. Semin Neurol 13: 256-263.
7. Koffman BM, Rugiero M, Dalakas MC (1998) Immune-mediated conditions and antibodies associated with sporadic inclusion body myositis. Muscle Nerve 21: 115-117.
8. Askanas V, Engel WK (1998) Sporadic inclusion-body myositis and its similarities to Alzheimer disease brain. Recent approaches to diagnosis and pathogenesis, and relation to aging. Scand J Rheumatol 27: 389-405.
9. Bilak M, Askanas V, Engel WK (1993) Strong immunoreactivity of alpha 1-antichymotrypsin co-localizes with beta-amyloid protein and ubiquitin in vacuolated muscle fibers of inclusion-body myositis. Acta Neuropathol 85: 378-382.
10. Askanas V, Bilak M, Engel WK et al. (1993) Prion protein is abnormally accumulated in inclusion-body myositis. Neuroreport 5: 25-28.
11. Roses AD (1996) Apolipoprotein E alleles as risk factors in Alzheimer's disease. Annu Rev Med 47: 387-400.
12. Sandbrink R, Hartmann T, Masters CL et al. (1996) Genes contributing to Alzheimer's disease. Mol Psychiatry 1:27-40.
13. Garlepp MJ, Tabarias H, van Bockxmeer FM et al. (1995) Apolipoprotein E epsilon 4 in inclusion body myositis. Ann Neurol 38: 957-959.
14. Lampe JB, Goßrau G, Gestrich B et al. (1999) Prion gene and apolipoprotein E in sporadic inclusion-body myositis. Ann Neurol 46: 485.
15. Harrington CR, Anderson JR, Chan KK (1995) Apolipoprotein E type epsilon 4 allele frequency is not increased in patients with sporadic inclusion-body myositis. Neurosci Lett 183: 35-38.
16. Goldfarb LG, Petersen RB, Tabaton M et al. (1992) Fatal familial insomnia and familial Creutzfeldt-Jakob disease: disease phenotype determined by a DNA polymorphism. Science 258: 806-808.
17. Laplanche JL, Delasnerie-Laupretre N, Brandel JP et al. (1994) Molecular genetics of prion diseases in France. French Research Group on Epidemiology of Human Spongiform Encephalopathies. Neurology 44: 2347-2351.
18. Deslys JP, Marce D, Dormont D (1994) Similar genetic susceptibility in iatrogenic and sporadic Creutzfeldt- Jakob disease. J Gen Virol 75: 23-27.
19. Palmer MS, Dryden AJ, Hughes JT et al. (1991) Homozygous prion protein genotype predisposes to sporadic Creutzfeldt- Jakob disease. Nature 352: 340-342.
20. Salvatore M, Genuardi M, Petraroli R et al. (1994) Polymorphisms of the prion protein gene in Italian patients with Creutzfeldt-Jakob disease. Hum Genet 94: 375-379.

21. Schulz-Schaeffer WJ, Giese A, Windl O et al. (1996) Polymorphism at codon 129 of the prion protein gene determines cerebellar pathology in Creutzfeldt-Jakob disease. Clin Neuropathol 15:353-357.
22. Hill AF, Desbruslais M, Joiner S et al. (1997) The same prion strain causes vCJD and BSE. Nature 389: 448-450.
23. Lampe J, Kitzler H, Walter MC et al. (1999) Methionine homozygosity at prion gene codon 129 may predispose to sporadic inclusion-body myositis. Lancet 353: 465-466.
24. Lampe JB, Goßrau G, Walter MC et al. (2000) Prion gene genetics and sporadic inclusion-body myositis. Alzheimer's Report 3 (Suppl. 1): S35.
25. Mendell JR, Sahenk Z, Gales T et al. (1991) Amyloid filaments in inclusion body myositis. Novel findings provide insight into nature of filaments. Arch Neurol 48: 1229-1234.
26. DeArmond SJ, Prusiner SB (1997) Prion diseases. In: Greenfield's Neuropathology (Graham DI, Lantos PL, eds), pp235-280, 6th ed. Vol II: Arnold, London.
27. Hsiao K.H, Baker F, Crow TJ et al. (1989) Linkage of a prion protein missense variant to Gerstmann-Straussler syndrome. Nature 338: 342-345.
28. Rosenthal NP, Keesey J, Crandall B et al. (1976) Familial neurological disease associated with spongiform encephalopathy. Arch Neurol 33: 252-259.
29. Beck EP, Daniel M, Parry HP (1964) Degeneration of the cerebellar and hypothalamo-neurohypophysial systems in sheep with scrapie; and its relationship to human system degenerations. Brain 87: 153-176.
30. Bosanquet FD, Daniel PM, Parry HB (1956) Myopthy in sheep. Its relationship to scrapie and to dermatomyositis and muscular dystrophy. Lancet ii: 737-746.
31. Hulland WH (1958) The skeletal muscle of sheep affected with scrapie. J Comp Pathol Ther 68: 264-274.
32. DeArmond SJ. Prusiner SB (1995) Etiology and pathogenesis of prion diseases. Am J Pathol 146: 785-811.
33. Global Surveillance, Diagnosis and Therapy of Human Transmissible Spongiform Encephalopathies. Report of a WHO Consultation and WHO/EMC/ZDI/98.9. 1998.
34. Sarkozi E, Askanas V, Engel WK (1994) Abnormal accumulation of prion protein mRNA in muscle fibers of patients with sporadic inclusion-body myositis and hereditary inclusion- body myopathy. Am J Pathol 145: 1280-1284.
35. Askanas V, Bilak M. Engel WK et al. (1993) Prion protein is strongly immunolocalized at the postsynaptic domain of human normal neuromuscular junctions. Neurosci Lett 159: 111-114.
36. Westaway D., DeArmond SJ, Cayetano-Canlas J et al. (1994) Degeneration of skeletal muscle, peripheral nerves, and the central nervous system in transgenic mice overexpressing wild-type prion proteins. Cell 76: 117-129.
37. Mitrani-Rosenbaum S, Argov Z, Blumenfeld A et al. (1996) Hereditary inclusion body myopathy maps to chromosome 9p1-q1. Hum Mol Genet 5: 159-163.
38. Martinsson T., Darin N, Kyllerman M et al. (1999) Dominant hereditary inclusion-body myopathy gene (IBM3) maps to chromosome region 17p13.1. Am J Hum Genet 64: 1420-1406.
39. Mikol J, Engel AG (1994) Inclusion body myositis. In: Myology (Engel AG, Franzini Armstrong C, eds), pp. 1384-1398, McGraw-Hill, New York.
40. Griggs RC, Askanas V, DiMauro S et al. (1995) Inclusion body myositis and myopathies. Ann Neurol 38: 705-713.
41. Hohlfeld R, Goebels N, Engel AG (1993) Cellular mechanisms in inflammatory myopathies. Baillières Clin Neurol 2: 617-635.

42. Fyhr IM, Moslemi AR, Mosavi AA et al. (1997) Oligoclonal expansion of muscle infiltrating T cells in inclusion body myositis. J Neuroimmunol 79: 185-189.
43. Hohlfeld R. Engel AG (1991) Coculture with autologous myotubes of cytotoxic T cells isolated from muscle in inflammatory myopathies. Ann Neurol 29: 498-507.
44. Garlepp MJ (1993) Immunogenetics of inflammatory myopathies. Baillières Clin Neurol 2: 579-297.
45. Garlepp MJ, Laing B, Zilko PJ (1994) HLA associations with inclusion body myositis. Clin Exp Immunol 98: 40-45.
46. Koffman BM, Sivakumar K, Simonis T et al. (1998) HLA allele distribution distinguishes sporadic inclusion body myositis from hereditary inclusion body myopathies. J Neuroimmunol 84: 139-142.
47. Rider LG, Gurley RC, Pandey JP et al. (1998) Clinical, serologic, and immunogenetic features of familial idiopathic inflammatory myopathy. Arthritis Rheum 41:710-719.
48. Lindberg C, Oldfors A, Tarkowski A (1995) Local T-cell proliferation and differentiation in inflammatory myopathies. Scand J Immunol 41: 421-426.
49. Kok CC, Croager EJ, Witt CS et al. (1999) Mapping of a candidate region for susceptibility to inclusion body myositis in the human major histocompatibility complex. Immunogenetics 49: 508-516.
50. Goßrau G, Füssel M, Walter MC et al. (2000) Histocompatibility antigens in sporadic inclusion-body myositis. Alzheimer's Report 3 (Suppl 1): S58.
51. Fyhr IM, Moslemi AR, Tarkowski A et al. (1996) Limited T-cell receptor V gene usage in inclusion body myositis. Scand J Immunol 43: 109-114.
52. Carpenter S, Karpati G, Heller I et al. (1978) Inclusion body myositis: a distinct variety of idiopathic inflammatory myopathy. Neurology 28: 8-17.
53. Chou SM (1986) Inclusion body myositis: a chronic persistent mumps myositis? Hum Pathol 17: 765-777.
54. Fox SA, Ward BK, Robbins PD et al. (1996) Inclusion body myositis: investigation of the mumps virus hypothesis by polymerase chain reaction. Muscle Nerve 19: 23-28.
55. Nishino H, Engel AG Rima BK. (1989) Inclusion body myositis: the mumps virus hypothesis. Ann. Neurol. 25:260-264.
56. Cupler EJ, Leon-Monzon M, Miller J et al. (1996) Inclusion body myositis in HIV-1 and HTLV-1 infected patients. Brain 119: 1887-1893.
57. Schlesinger I, Soffer D, Lossos A et al. (1996) Inclusion body myositis: atypical clinical presentations. Eur Neurol 36: 89-93.
58. Abarbanel JM, Lichtenfeld Y, Zirkin H et al. (1988) Inclusion body myositis in post-poliomyelitis muscular atrophy. Acta Neurol Scand 78: 81-84.
59. Gajdusek DC (1996) Infectious amyloidosis: subacute spongiform encphalopathies as transmissible cerebral amyloidosis. In: Fields Virology (Fields BN, Knipe DM, Howley PM eds), pp. 2851-2900, Lippincott-Raven Publishers, Philadelphia.
60. Come JH, Fraser PE, Lansbury PT Jr. (1993) A kinetic model for amyloid formation in the prion diseases: importance of seeding. Proc Natl Acad Sci USA 90: 5959-5963.
61. Welch WJ, Gambetti P (1998) Chaperoning brain diseases. Nature 392: 23-24.
62. Banwell BL, Engel AG (2000) AlphaB-crystallin immunolocalization yields new insights into inclusion body myositis. Neurology 54:1033-1041.
63. Askanas V, Engel WK, Alvarez RB et al. (2000) Inclusion body myositis, muscle blood vessel and cardiac amyloidosis, and transthyretin Val122Ile allele. Ann. Neurol. 47:544-549.
64. Askanas V, McFerrin J, Alvarez RB (1997) Beta APP gene transfer into cultured human muscle induces inclusion- body myositis aspects. Neuroreport 8: 2155-2158.

65. Fukuchi K, Pham D, Hart M et al. (1998) Amyloid-beta deposition in skeletal muscle of transgenic mice: possible model of inclusion body myopathy. Am J Pathol 153: 1687-1693.
66. Jin LW, Hearn MG, Ogburn CE et al. (1998) Transgenic mice over-expressing the C-99 fragment of betaPP with an alpha-secretase site mutation develop a myopathy similar to human inclusion body myositis. Am J Pathol 153: 1679-1686.
67. Yang CC, Alvarez RB, Engel, WK et al. (1996) Increase of nitric oxide synthases and nitrotyrosine in inclusion-body myositis. Neuroreport 8: 153-158.
68. Askanas V, Engel WK (1998) Sporadic inclusion-body myositis and hereditary inclusion-body myopathies: diseases of oxidative stress and aging? Arch Neurol 55: 915-920.
69. Yang CC, Alvarez RB, Engel WK (1998) Nitric oxide-induced oxidative stress in autosomal recessive and dominant inclusion-body myopathies. Brain 121: 1089-1097.
70. Schroder JM, Molnar M (1997) Mitochondrial abnormalities and peripheral neuropathy in inflammatory myopathy, especially inclusion body myositis. Mol Cell Biochem 174: 277-281.
71. Askanas V, Engel WK, Alvarez RB (1992) Light and electron microscopic localization of beta-amyloid protein in muscle biopsies of patients with inclusion-body myositis. Am J Pathol 141:31-36.

MITOCHONDRIAL DYSFUNCTION IN NEURODEGENERATIVE DISORDERS AND AGEING

[1]Christopher Turner and [1,2]Anthony H.V. Schapira
[1]University Department of Clinical Neurosciences, Royal Free and University College Medical School and [2]Institute of Neurology, University College London, London NW3 2PF, UK

1. INTRODUCTION: THE SPECTRUM OF MITOCHONDRIAL FUNCTION

In eukaryotic cells, mitochondria are the organelles that produce the majority of adenosine triphosphate (ATP) required for normal neuronal function and survival. ATP is generated by oxidative phosphorylation (OXPHOS) within mitochondria from intermediates, such as NADH and $FADH_2$ which are produced by ß-oxidation and the Kreb's cycle.

Mitochondria also have a pivotal role in the induction of apoptosis by the release of pro-apoptotic proteins. The competence of mitochondria to produce ATP will determine whether cell death occurs by apoptosis or necrosis. Chronic mitochondrial dysfunction could impair neuronal function and lead to neurodegeneration. Mitochondria have several other roles in maintaining cellular homeostasis such as storage of intracellular calcium, fatty acid oxidation, the Kreb's cycle and iron metabolism.

This article will critically review the evidence for the role of mitochondria in neurodegenerative diseases and the ageing brain.

Neuropathology and Genetics of Dementia, Edited by Tolnay and Probst
Kluwer Academic/Plenum Publishers, New York 2001

2. MITOCHONDRIAL CELL BIOLOGY

The bacterial hypothesis of the origin of mitochondria suggests that α-purple bacteria were incorporated into eukaryotic cells and, during evolution, these bacteria transferred many of their essential genes to the nuclear chromosomes[1]. The mitochondria still have remnants of their bacterial origin such as the use of N-formylmethionyl-tRNA as the initiator of protein synthesis[2,3].

The mitochondrial respiratory chain and oxidative phosphorylation system comprise five multisubunit protein complexes, which are located in the inner mitochondrial membrane, and two mobile electron carriers, cytochrome c and ubiquinone (Figure 1).

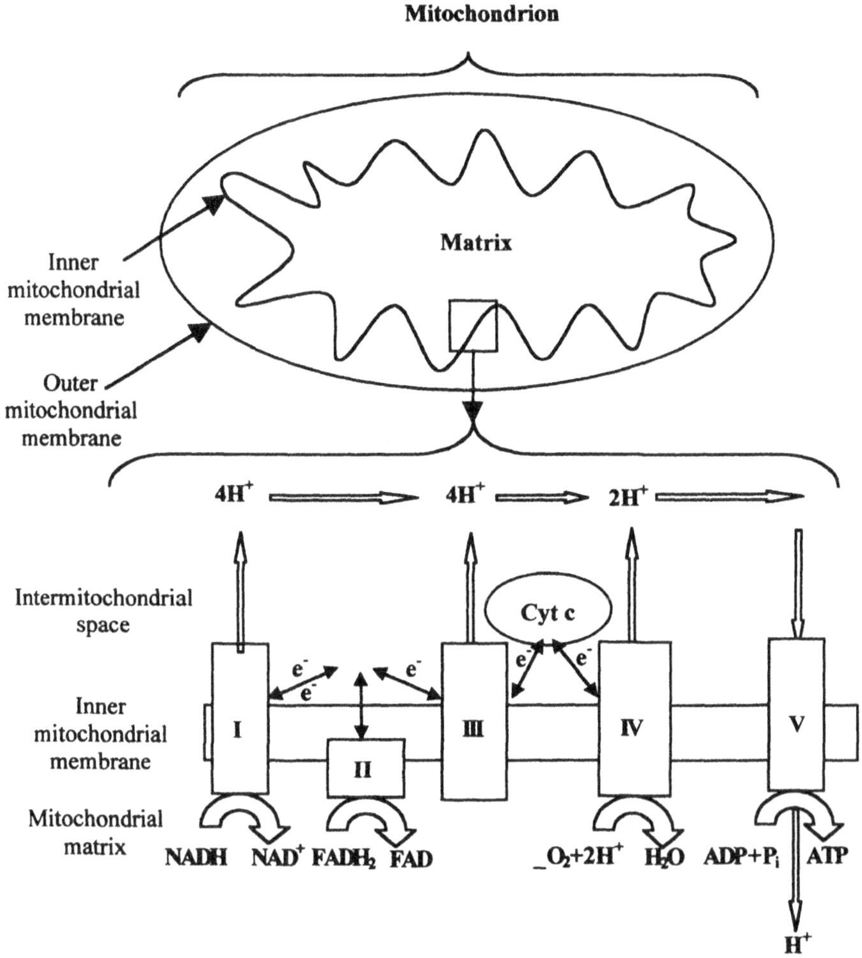

Figure 1. A schematic diagram of the mitochondrial respiratory chain. Cyt c=cytochrome c; Q=ubiquinone

The OXPHOS system produces ATP by the production of a proton gradient across the inner mitochondrial membrane which drives ATP production by ATP synthase or complex V. The proton gradient is produced by the release of protons from complexes I, III and IV into the intermembranous space. The energy for this process is yielded from the transfer of electrons from NADH/$FADH_2$ through the protein complexes to the final electron acceptor, oxygen. Complex V couples the re-entry of protons to the mitochondrial matrix with the production of ATP. Complex III and to a lesser extent complex I are major sites of generation of potentially toxic reactive oxygen species (ROS)[4].

Each mitochondrion contains 2-10 molecules of circular double-stranded mitochondrial DNA (mtDNA), which encodes for 22 transfer RNAs, two ribosomal RNAs and 13 proteins: seven subunits of complex I, cytochrome b of complex III, COI, COII and COIII of complex IV (cytochrome oxidase) and subunits 6 and 8 of complex V (ATPase)[5,6]. MtDNA is inherited through the maternal line. Greater than 80 mtDNA mutations have been associated with a wide spectrum of human disease[7]. The mutations usually co-exist with normal wild-type molecules i.e. heteroplasmy[8]. The proportion of the mutant form varies between tissues depending on segregation during mitosis. A high level of the mutant form is required to produce respiratory chain dysfunction, although this is partly determined by the level of dependence of the tissue on oxidative metabolism.

Neurons require a high level of ATP production to restore ionic homeostasis following the controlled flux of ions across the cell membrane during electrical signalling. Most of the ATP utilised by Na^+/K^+ ATPase and Ca^{2+} ATPase in maintaining ionic homeostasis is generated in mitochondria.

3. THE MITOCHONDRIAL HYPOTHESIS IN NEURODEGENERATION

This hypothesis postulates that defects in mitochondrial metabolism may lead to a chronic depletion in cellular ATP. This may impair neuronal function and depending on the severity of the ATP deficiency trigger the neuron to enter apoptosis. Impaired OXPHOS function may be a consequence of mutations of genes (nuclear or mitochondrial) encoding subunits of complexes I-V, nuclear gene mutations of proteins involved in the assembly or stability of OXPHOS proteins, or endogenous and exogenous mitochondrial toxins (Figure 2)[9,10]. This hypothesis is supported by evidence that disorders of mitochondrial function caused by mutations in mitochondrial proteins/nucleic acids or mitochondrial inhibitors have a role in a variety of neurodegenerative diseases.

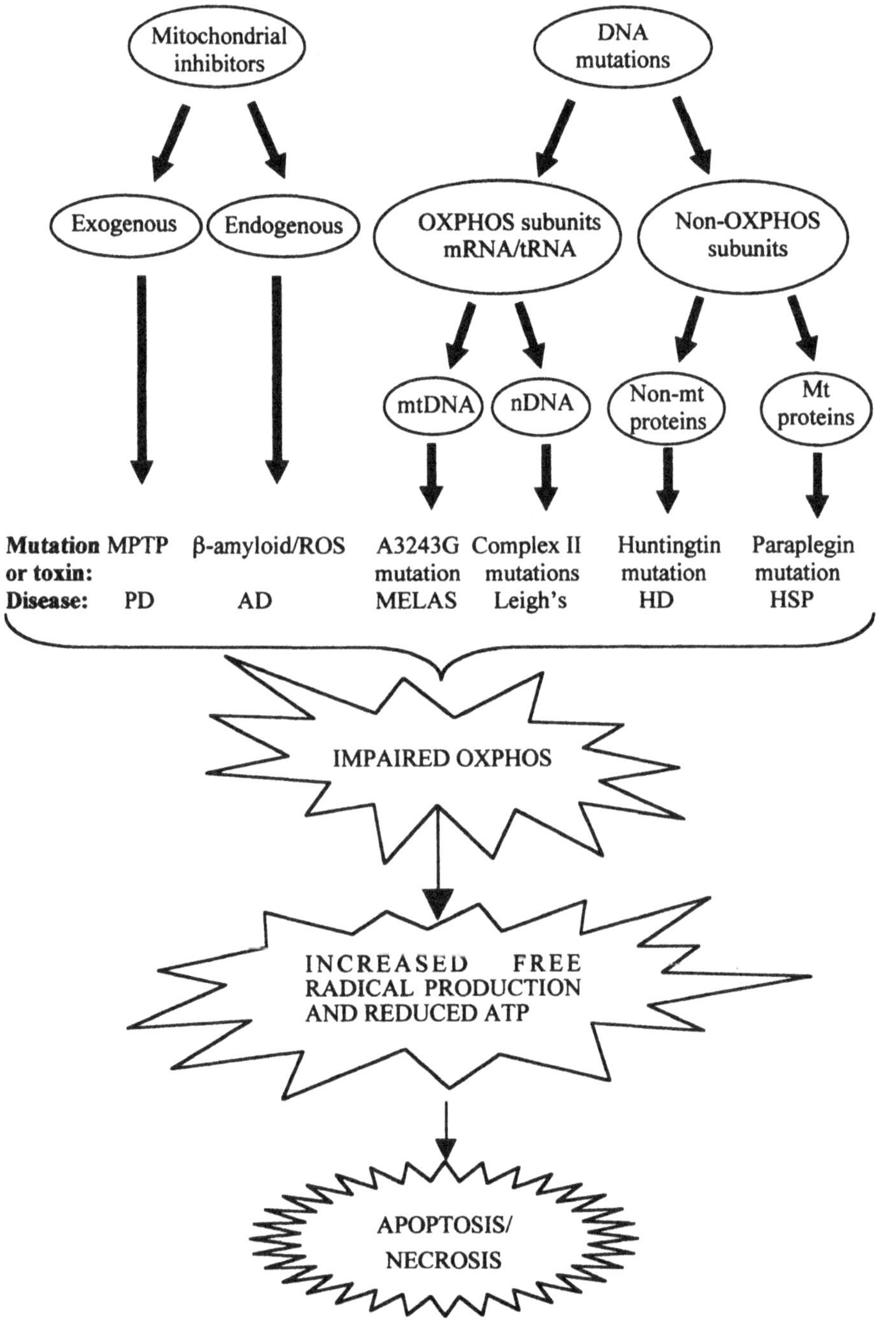

Figure 2. The mitochondrial hypothesis of neurodegeneration

Normal ageing may be associated with a gradual reduction in mitochondrial function secondary to the accumulation of mtDNA mutations or other mitochondrial inhibitory processes. Discussion is limited to the role of mitochondrial dysfunction in several neurodegenerative diseases including Parkinson's disease (PD), Huntington's disease (HD), Alzheimer's disease (AD), Friedreich's ataxia (FA) and hereditary spastic paraplegia (HSP) as well as the ageing brain.

4. PARKINSON'S DISEASE

PD is a common neurodegenerative disease associated with tremor, bradykinesia, rigidity and loss of postural reflexes. The predominant pathology is loss of dopaminergic neurons in the substantia nigra (SN) pars compacta. Many factors have been implicated in the pathogenesis of PD including genetic predisposition, increased deposition of iron in the SN, enhanced oxidative stress due to monoamine oxidase activity and respiratory chain inhibition, and weak excitotoxicity due to chronic mitochondrial dysfunction.

A specific complex I deficiency limited to the substantia nigra was first described in 1989[11,12]. A similar defect has not been described in multiple system atrophy (MSA) suggesting that neurodegeneration and exposure to L-dopa are not in themselves associated with a complex I deficiency[13]. A reduction in complex I activity has not been found in the cingulate gyrus of CLBD or AD suggesting that the presence of Lewy bodies and gliosis do also not cause the complex I deficiency[14]. The dopaminergic neurons in the SN represent less than 5% of the total cell population in this region and therefore the 35% reduction in complex I activity must reflect involvement of both glia and neurons. However, dopaminergic neurons are highly dependent on aerobic metabolism which renders these cells particularly vulnerable to mitochondrial dysfunction.

Evidence of respiratory chain dysfunction outside the CNS has been investigated extensively. The application of ^{31}phosphorus magnetic resonance spectroscopy to PD skeletal muscle have demonstrated OXPHOS defects in some studies but not in others[15-17]. Similarly studies of skeletal muscle homogenates demonstrate occasional severe defects, but the majority of patients show no abnormality[18].

In contrast, there is clear evidence that there is mitochondrial dysfunction in PD platelets. Most studies using cybrids demonstrate defects in complex I but also in II-IV[18,19]. Cybrids are cells whose mitochondria have been removed (ρ^0 cells) by exposure over several passages to agents such as ethidium bromide and fused with an enucleated cell or platelets, which contain mitochondria and mtDNA[19,20]. These cybrids contain mtDNA and mtDNA-encoded respiratory chain subunits from the donor and nuclear encoded subunits from the host ρ^0 cells. In this model, if a defect in mitochondrial function is perpetuated from the donor mitochondria to the cybrids then it will be due to a defect in the donor mtDNA.

Platelets from PD patients with the lowest complex I activity fused with A549 ρ^0 cells demonstrated a 25% reduction in complex I activity suggesting that there was a defect in host PD mtDNA[19]. Further investigation suggested that the mtDNA defect is heteroplasmic. The cybrid system has also demonstrated a complex I defect in unselected PD patients[20].

These findings suggest that there may be a mutation in one of the mtDNA complex I genes in PD. The cybrid experiments do not distinguish between somatic and inherited mutations. Most cases of PD are sporadic and do not follow a maternal pattern of inheritance. Somatic mtDNA mutations may be secondary to a toxic insult on the patient's haemopoietic system. Most cases of mitochondrial myopathy are sporadic e.g. 40% of Leber's hereditary optic neuropathy, and therefore a history of maternal inheritance is not necessary for mtDNA involvement. However, a consistent mutation in PD mtDNA has not been found. A higher incidence of some polymorphisms in mtDNA have been described in PD than controls but these probably reflect a difference in ethnicity and haplotype background. The investigation of a possible consistent mtDNA base change which continues through the maternal lineage and PD nuclear backgrounds in the cybrid model will hopefully clarify the origin of the complex I mutation.

4.1 Environmental factors

The agent 1-methyl-4-phenyl-1,2,3,6-tetrahydropyridine (MPTP) was identified as a contaminant of a meperidine analogue "designer drug" responsible for inducing parkinsonism in a group of Californian drug abusers.

MPTP induces clinical features similar, but not identical, to idiopathic PD. A pathological study of a brain from a patient exposed to MPTP who developed parkinsonism demonstrated severe destruction of the dopaminergic neurons of the substantia nigra[21]. ^{18}-fluoro-dopa positron emission tomography in patients with MPTP-induced parkinsonism indicated progression of nigrostriatal loss over seven years at a rate faster than ageing and comparable with idiopathic PD[22]. This suggests toxicity from the compound long after initial exposure and suggests ongoing biochemical dysfunction culminating in cell death. MPTP is a protoxin metabolised to its active derivative 1-methyl-4-phenylpyridinium (MMP+) by monoamine oxidase B (MAO-B). MPP^+ is a specific reversible inhibitor of complex I and probably interacts with complex I at the same site as rotenone and piericidin A[23]. MPP^+ induces a more severe and irreversible inhibition of complex I if cytochrome oxidase (complex IV) is inhibited[24]. This can be prevented with free radical scavengers indicating that complex I inhibition is partly mediated by oxidative stress. The inhibition of complex I leads to increased free radical production and hence a self-amplifying cycle of complex I inhibition and damage may result in progressive cell damage. This may explain the latent effect of exposure to MPTP in humans.

MPP+ may also generate free radical production independently of complex I inhibition[25,26]. The nitric oxide synthase inhibitor, 7-nitroindazole, protects monkeys and rodents from MPTP[27,28]. However there is also evidence that 7-nitroindazole inhibits MAO-B which would prevent conversion of MPTP to MPP^+, although toxicity from MPP^+ is also prevented by nitric oxide synthase inhibitors[29].

Physiological correlates of MPTP have been described: 1,2,3,4-tetrahydroisoquinoline (TIQ) is structurally related to MPTP and has been found in human brain. There is some evidence that it is found in greater quantities in PD brain, that it is toxic to dopaminergic neurons and can induce motor deficits in monkeys[30,31]. TIQ is also a complex I inhibitor[32] and its methylation product *N*-methyl TIQ is a substrate for MAO-B producing *n*-methyl isoquinolinium (NMIQ+) which inhibits tyrosine hydroxylase and MAO. This compound is also found in certain foodstuffs, can cross the blood-brain barrier and can be formed by condensation reaction in the brain.

5. HUNTINGTON'S DISEASE

HD is characterised by a movement disorder and dementia. It is an autosomal dominant condition caused by an abnormal CAG expansion within the IT15 gene on chromosome 4 which encodes huntingtin a 349kDa protein of unknown function[33]. The neuropathology is widespread in the HD brain, but is especially severe in the caudate nucleus where the medium GABA-ergic spiny neurons are mostly affected. The putamen and cortex are less affected and the cerebellum is relatively spared[34]. As in PD the cause of such specific neuronal loss is uncertain.

Slow excitotoxicity may have a significant role in the pathogenesis of HD. A critical factor in this mechanism is a defect in energy metabolism which releases the magnesium blockade of NMDA receptor activation and renders ambient levels of glutamate excitotoxic. This process also involves activation of nitric oxide synthase and the production of nitric oxide.

There is in vivo and post-mortem evidence for dysfunction of energy metabolism in HD. Striatal and cerebral cortex glucose metabolism is decreased in HD and precedes bulk tissue loss[35-37]. Using magnetic resonance spectroscopy, elevated lactate levels in the occipital cortex have been found to correlate with the length of the CAG repeat in HD striatum[38]. Other studies have not verified these findings[39]. The lactate/pyruvate ration has been found to be mildly but consistently elevated in HD and another trinucleotide repeat neurodegenerative disease, Machado-Joseph. In the same study, ^{31}P MRS was performed on HD muscle and demonstrated a decreased ratio of phosphocreatine to inorganic phosphats[40]. Currently, ubiquinone, an essential component of the respiratory chain, is undergoing clinical trials following the demonstration of a reduction in cortical lactate levels following ubiquinone administration in HD[40].

Direct measurement of respiratory chain activity in HD caudate demonstrated a severe deficiency of complex II and III (56%) and IV (33%) activity[41]. A similar but less severe pattern was observed in HD putamen but not cortex or cerebellum[42,43]. Aconitase is an iron-sulphur (FeS) containing enzyme which is involved in the Kreb's cycle and iron homeo-stasis. Its activity is especially susceptible to inhibition by $O_2^{\cdot -}$ and by the reaction product of $O_2^{\cdot -}$ with $NO^{\cdot}$, $ONOO^{-}$ or peroxynitrate[44-46]. Aconitase deficiency has been found in HD caudate (92%), putamen (73%) and cortex (48%) but not cerebellum and this deficiency closely follows the pathology of HD[43]. Complexes II and III are FeS-containing compounds and are susceptible to inhibition by these free radicals. The pattern of enzyme loss in HD suggests that free radicals and excitotoxity have a role in the pathogenesis of the disease.

There are conflicting results regarding respiratory chain activity in HD platelets: some demonstrate a defect in complex I activity whilst others are normal[41,47]. The activities in HD fibroblasts have been found to be normal[43].

Skeletal muscle offers a tissue that can be relatively easily biopsied and, similar to brain, has high dependence on oxidative metabolism for the generation of ATP. A complex I deficiency has been described in 3 out of 4 muscle biopsies on symptomatic HD patients[48]. This implies that there is a direct relationship between the mitochondrial defect and the expanded CAG repeat in HD. We are currently studying respiratory chain activity in muscle from pre-symptomatic and symptomatic HD patients[49,50].

5.1 Models of Huntington's disease

The systemic administration of 3-nitropropionic acid in rats and primates is used to produce animal models displaying lesions that closely resemble the neuropathological features of HD. This model produces metabolic impairment and complex II inhibition which is age-dependent, in that younger animals are more resistant to the neurotoxic effects[51].

Several transgenic mouse models of HD have been created. The first to demonstrate a clinical and pathological phenotype similar to HD was produced with a transgene containing one kilobase of the human HD promoter region, exon one carrying CAG repeat expansions of 141-157 units, and 262 base pairs of intron one[52]. This R6/2 transgenic mouse exhibits progressive overt neurological disease from 2 months of age. Both the light microscopic and ultrastructural pathology are very similar to those seen in HD brain. These neuropathological features occur approximately 4 weeks prior to a progressive movement disorder and muscle wasting and 10 weeks before neuronal cell death in selective brain regions. This suggests that neuronal dysfunction is responsible for the initial clinical phenotype rather than cell death.

Several defects in the respiratory chain have recently been characterised in the R6/2 mouse model. A reduction in complex IV in the striatum and cerebral cortex and a reduction in aconitase in the striatum has recently been described[53]. These were associated with increased immunostaining for inducible nitric oxide synthase (iNOS) and nitrotyrosine (a marker of increased peroxynitrate generation) in the mouse brains. There is further evidence for excitotoxicity and free radical damage in these transgenic mice from experiments demonstrating a loss of presynaptic metabotropic glutamate receptor 2 (mGluR2) which regulate glutamate release from corticostriatal neurons[54]. Further evidence demonstrated increased glutamate receptor sensitivity to NMDA in these mice[55]. This may lead to dysregulated glutamate release and excitotoxic cell dysfunction/death. An increase in the lesion size produced by 3-nitropropionic acid in the R6/2 mice and increased striatal 3,4-dihydroxybenzoic acid (a marker of ROS) also support a role for mitochondrial dysfunction and free radical damage[56].

In summary, defects in energy metabolism and respiratory chain activity associated with excitotoxicity and increased free radical production are now well established in HD. The transgenic mouse models of HD are enabling the dissection of the sequence of molecular events from CAG expansion to neuronal cell dysfunction and death. The mechanism by which a non-mitochondrial protein, huntingtin, exerts its effect on energy metabolism and mitochondrial function remain elusive and will be the source of future research.

6. HEREDITARY SPASTIC PARAPLEGIA

Autosomal recessive, dominant and X-linked forms of HSP have been described. Patients manifest from childhood to adulthood with progressive leg stiffness and weakness as well as neuropathy, ataxia, retinitis, optic atrophy and deafness in a proportion of "HSP plus" patients.

A new gene defect in pure HSP families with autosomal recessive inheritance has been described linked to 16q24.3. The gene product is a nuclear-encoded mitochondrial non-respiratory chain protein of 795 amino acids called paraplegin[57]. Paraplegin contains an N-terminus that targets it to the mitochondria but its function is unknown.

Muscle biopsies from four patients with paraplegin mutations demonstrated mitochondrial abnormalities. The most severe cases had ragged red, succinate dehydrogenase (complex II) positive fibres which stained negative for cytochrome oxidase (complex IV) and contained paracrystalline inclusions. These features are typical of myopathies associated with mtDNA mutations.

Future studies will define the biochemical changes associated with this mitochondrial defect and the pathological role of a nuclear-encoded non-respiratory chain protein, paraplegin in defective mitochondrial function.

7. FRIEDREICH'S ATAXIA

FA is an autosomal recessive disorder characterised by progressive ataxia, dysarthria, skeletal deformities, neuropathy, pyramidal features and a hypertrophic cardiomyopathy. Most patients are wheelchair-bound within 10-15 years of onset in adolescence and die from progressive heart failure.

The pathology is primarily a "dying-back" axonopathy of the CNS and PNS affecting large sensory axons of the dorsal root ganglia (DRG) and the spinocerebellar and pyramidal tracts in the spinal cord.

The molecular pathology in 98% of cases has been identified as an expanded and unstable GAA triplet repeat in intron 1 of the frataxin gene (X25) resulting in the deficiency of frataxin protein[58,59]. The expression of frataxin is highest in the spinal cord, DRG and heart which correlates well with the pathology.

Frataxin contains a predicted N-terminal mitochondrial targeting sequence and fluorescence microscopy has confirmed the mitochondrial localisation of frataxin[60,61]. The function of frataxin is unknown. A deletion of the yeast frataxin homologue,YFH1, results in an inability to grow on glycerol and ethanol suggesting a defect in oxidative phosphorylation. The deletion of YFH1 also results in damage or loss of mtDNA suggesting that frataxin may be involved in mtDNA maintenance or replication[62], but this may be a non-specific effect of the yeast model[63]. Frataxin may also have a role in mitochondrial iron metabolism. In yeast cells with a deletion of YFH1 gene there is continued expression of the high affinity iron transport system even in the presence of iron rich medium. In this mutant there are high levels of iron in the mitochondria unrelated to cellular iron levels which may lead to elevated levels of ROS in close association with mtDNA. The subsequent elevated levels of mutant mtDNA or the direct effect of ROS on respiratory chain proteins could lead to defects in energy metabolism. Iron accumulation has also been demonstrated in FA myocardium implicating frataxin in generalised intramitochondrial iron metabolism[64]. The similarity between vitamin E deficiency and FA has suggested that frataxin may be involved in the antioxidant system which would support the role of oxidative damage in FA[59]. Vitamin E deficiency in rats also results in decreased complex I-IV activities[65].

Serum lactate levels are elevated in FA, which suggests defective oxidative phosphorylation[66]. The initial studies into individual mitochondrial enzyme activities in FA produced some conflicting data. Recently two studies have suggested severe deficiencies in the activities of complexes I-III and aconitase[67,68]. All these proteins contain iron-sulphur clusters making them sensitive to free radical damage.

Recent MRS studies have demonstrated a correlation between GAA repeat length and defects in OXPHOS[69]. This suggests that frataxin causes an OXPHOS defect and further supports a role for mitochondrial dysfunction in FA.

The current evidence in FA pathogenesis points to a role of oxidative stress, especially in mitochondria, associated with defects in the respiratory chain. This has precipitated trials of antioxidants which are currently under investigation.

8. ALZHEIMER'S DISEASE

AD is a progressive dementia affecting 5-15% of people over 65[70]. The pathology mostly involves the hippocampus and neocortex, especially the temporal lobes where the neuropathological hallmarks of neurofibrillary tangles (NFT) and neuritic ß-amyloid plaques (NP) are found. The main clinical feature is progressive memory loss. A small proportion of patients with AD have an autosomal dominant form associated with mutations in the amyloid precursor protein (APP) and presenilin genes. Most patients with AD have sporadic disease and the aetiology and pathogenesis remain elusive.

8.1 Mitochondrial dysfunction and AD

An alteration in energy metabolism is an early and prominent feature of AD brain[71]. Significant decreases in oxygen and glucose utilisation have been demonstrated in PET studies as the disease progresses[72]. An early defect in glucose metabolism is seen in the parietal and temporal regions, which are areas particularly associated with increased densities of neuritic plaques[73].

Further direct evidence for a defect in mitochondrial function comes from studies of respiratory chain activity. Mitochondria prepared from fresh AD hemi-brains have demonstrated a 40% reduction in complex I and a 53% reduction in COX (complex IV) activity[74,75]. When homogenates of AD brain and not extracted mitochondria are tested there was a significant deficit in COX activity in the temporal and parietal regions[76]. Several further studies, including some that corrected for citrate synthase activity, have confirmed the defect in COX activity[77,78].

Histochemical data have demonstrated a defect in COX activity in the molecular layer of the dentate gyrus and other parts of the hippocampus[79]. The perforant pathway between the entorhinal cortex and the outer molecular layer of the dentate gyrus is one of the most severely affected in AD.

In situ hybridization studies have demonstrated decreased mRNA levels of the mtDNA-encoded subunit II but not the nDNA-encoded subunit IV of COX in AD hippocampus[80]. However, other histochemical studies have demonstrated that the defect is not confined to mitochondrial-encoded subunits and the pattern of reduced expression of both mtDNA and nDNA-encoded genes are consistent with a physiological down-regulation of the mitochondrial respiratory chain in response to reduced neuronal activity[81].

Studies in monocular deprivation of monkeys has demonstrated decreased mRNA of both nuclear and mitochondrial subunits of COX in the lateral geniculate nucleus (LGN) suggesting that deafferentation in itself may lead to defects in the respiratory chain[82].

8.2 Mitochondrial DNA mutations in AD

There is increasing evidence for mitochondrial dysfunction in AD but the pathological role of mtDNA mutations remains uncertain.

There has been one report of increased levels of a somatic cell i.e. acquired, mtDNA mutation, which is known as the "common deletion", in the cerebral cortex of AD but this has not been substantiated by most studies[83-85].

There is epidemiological evidence which suggests that there is a higher incidence of AD in mothers rather than fathers of probands[86]. This suggests a role for maternally inherited mtDNA mutations in AD.

There are conflicting data to support the role of the A4336G polymorphism in mtDNA in AD. Some studies suggest a frequency of 5.2% in AD and/or PD versus 0.7% in controls[87]. Other studies suggest a lower incidence of this polymorphism[88] and a "founder effect" has been implicated to explain some of the segregation[89].

In 1997 six novel mtDNA point mutations in COX (three in COX1 and three in COX2) were described in platelet-enriched DNA from late-onset AD patients[90]. These were transferred into a cybrid system which suggested biochemical abnormalites such as a decrease in COX activity[91]. These mutations were subsequently demonstrated to be due to an artefactual amplification of nucleus-embedded mtDNA pseudogenes[92].

In summary, there is evidence of impaired respiratory chain function in AD especially with regard to a deficiency of COX but these effects may be downstream of the initial pathological events and secondary to deafferentation. There is no convincing evidence to implicate sporadic mitochondrial mutations in the pathogenesis of AD but maternally-inherited polymorphisms may have a role in some cases.

8.3 Oxidative stress in AD

Free radical-mediated cellular toxicity has been one of many hypotheses to explain the pathogenesis of AD. The brain may preferentially be affected by oxidative stress because of its high oxygen consumption, high lipid content, the rapid turnover of mtDNA, and a lack of antioxidant defences especially within the mitochondria where the majority of cellular ROS are produced[93]. To support this hypothesis there are several studies that have demonstrated increased protein oxidation and lipid peroxidation in AD brains[94,95]. NFT have been demonstrated to contain protein carbonyls from protein oxidation and peroxynitrate[96,97]. Both NFT and NP have been demonstrated to contain advanced glycation end-products which are capable of producing free radicals[98,99]. The activity of an important anti-oxidant enzyme, catalase, is reduced in AD brains[100].

The physiological role of APP and the pathological role of one of its cleavage products found in the amyloid plaques in AD, ß-amyloid, still remain uncertain. There is some evidence that ß-amyloid impairs mitochondrial function. This has been studied most extensively in inclusion body myositis (IBM), which is a common myopathy in patients over the age of 50. IBM is characterised histologically by the presence of vacuolated muscle fibres showing accumulation of APP and paired helical filaments containing hyperphosphorylated tau[101]. The transfer of cDNA encoding APP751 into human myoblasts produces abnormalities of mitochondrial structure and function. A decrease in staining for cytochrome c oxidase (COX) is seen after 24 hours and is present in 80% of fibres by 2 weeks. Ultrastructural studies at 3-4 weeks revealed abnormal mitochondrial structure including the presence of paracrystalline inclusions[101].

There is also evidence of mtDNA oxidative damage by assessment of the oxidised radical adduct 8-hydroxy-2'-deoxyguanosine (OH^8dG) which increases with age in human muscle and brain[102]. Studies in AD brains demonstrate preferential accumulation in mtDNA versus nDNA and significantly higher levels than in controls[103].

9. MITOCHONDRIA AND AGEING

There is good evidence for the pathological role of the accumulation of nuclear DNA mutations with increasing age[104,105]. The contribution of mtDNA mutations to age-related decline in cellular function in humans and other mammals has also been postulated[106-108]. There is now growing evidence that mtDNA accumulates mutations with age. In normal aged individuals, mtDNA mutations accumulate preferentially in post-mitotic tissues with high oxidative requirements such as muscle and brain[109]. The substantia nigra (SN) accumulates mutations at higher levels than the rest of the brain[110]. It is postulated that the preferential targeting of neurons for mtDNA mutations is secondary to increased mitochondrial oxidative stress. Some of the deletion mutations are identical with those that occur in mitochondrial genetic diseases, such as Kearns-Sayre[111].

Kearns-Sayre syndrome (KSS) is characterised by paralysis of extraocular muscles, pigmentary retinopathy and cerebellar ataxia. It is associated with sporadic mtDNA rearrangements which are mostly large deletions (up to 9kB in size) and are present in up to 80% of mtDNA i.e. patients are heteroplasmic. The size, location and degree of heteroplasmy varies amongst patients. One particular deletion of 4977bp has been described in 33% of patients and is called the "common deletion"[112,113]. The mutant DNA is transcribed but not translated because of the disruption to essential mitochondrial tRNA which results in loss of translation of mRNA from genes outside of the deletion[114]. The "common deletion" accumulates exponentially in ageing human muscle by a factor of 10,000 over the course of the normal human lifespan, reaching 0.1% of total muscle mtDNA by 84 years[115]. There is focal accumulation of mtDNA deletions within the brain especially in the striatum, cerebral cortex and SN but not in the cerebellum[107,110].

The common deletion is often used as a biomarker of mitochondrial mutation but other deletions and rearrangements have been described in mtDNA from aged tissues (for review see[116]). The presence of mtDNA point mutations is less well substantiated[117,118].

The pathological role of somatic mtDNA mutations in the ageing process is uncertain, especially in view of their low levels in tissues. The clustering of groups of dysfunctional cells with many mtDNA mutations may account for the low levels in tissue homogenates[119].

The biochemical analysis of mitochondrial function with ageing has been investigated by three methods. Firstly, histochemical studies suggest that there is an exponential rise in COX-negative muscle fibres over the age of 40 years[120]. Secondly, the enzymatic activity of respiratory chain components decreases with age[121] in most studies but this has not been substantiated by all investigators[122]. Lastly, the mitochondrial membrane potential decreases with age in several species[123,124].

In summary, there is substantial evidence for mtDNA deletion accumulation with age and probably a deterioration in respiratory chain activity. Future research will need to try to link these defects with the phenotypic changes of ageing tissues.

10. MITOCHONDRIA: APOPTOSIS AND EXCITOTOXICITY

Apoptosis is a form of cell death characterised by an organised and controlled sequence of molecular and cellular events culminating in the removal of a cell from its potential or current physiological or pathological function. The mitochondrion has a critical role in this sequence of events in some forms of apoptosis[125].

Apoptosis is an ATP-requiring pathway and if cellular ATP falls too low during apoptotic stimulation then necrosis ensues[126]. Several proteins regulate apoptosis via an interaction with the mitochondrial permeability transition pore (PTP) or megapore[127]. The megapore spans both the inner and outer mitochondrial membrane and includes porin or the voltage-dependent anion channel (VDAC), the adenine nucleotide translocator, the peripheral benzodiazepine receptor, and cyclophilin D.

Complexes I, III and IV transfer protons from the inner mitochondrial membrane to the intermembranous space creating a potential difference across the inner membrane. A fall in potential opens the pore (permeability transition) and is associated with the release of pro-apoptotic molecules from the mitochondrion through the pore, including apoptosis-initiating factors (AIF) and the intermembranous OXPHOS protein cytochrome c. The release of these factors may also occur by rupture of the outer mitochondrial membrane. Several mitochondrial proteins regulate the opening of the VDAC such as Bax and Bak. These proteins bind to the pore and accelerate the release of cytochrome c into the cytosol. Another regulatory protein Bcl-X_L inhibits apoptosis by binding to and closing the pore preventing the release of cytochrome c[128]. In a VDAC-deficient yeast model the proapoptotic effects of Bak and Bax are lost suggesting that the VDAC is critical for mediating the action of these proteins. The megapore is also opened by high intramitochondrial calcium levels and by atractyloside. It is closed by low calcium levels and benzodiazepine agonists.

Once cytochrome c has been released from the mitochondrion it binds to apoptosis-activating factor-1 (Apaf-1) and cleaves and activates caspase 9, which in turn activates caspase 3[129-130]. The caspase cascade initiates the activation of further proteins and terminates in apoptotic cell death. AIF appears to act further downstream from the caspase cascade. AIF is capable of causing nuclear chromatin condensation and is insensitive to inhibitors of caspase-3[131].

There is good evidence in acute neurological injuries, such as stroke, that glutamate released during the insult activates the NMDA receptor leading to calcium influx and neuronal death[132]. The initiation of cell death probably involves accumulation of mitochondrial calcium[133]. This leads to increased generation of ROS and opening of the mitochondrial megapore. It is postulated that these events trigger both apoptotic and necrotic cell death under different environmental conditions. In stroke, the calcium influx is acute and massive and leads to necrosis. The role of excitotoxicity is less clear in neurodegenerative diseases. The "slow excitotoxic" hypothesis suggests that chronic depletion of ATP precipitated by defects in respiratory chain function results in enhanced production of ROS. This releases the magnesium-dependant blockade of the NMDA receptor and renders ambient levels of glutamate excitotoxic by enabling a chronic influx and accumulation of calcium within the mitochondria. This gradual accumulation of calcium results in apoptosis rather than necrosis. However, there is recent evidence that the R6/1 transgenic mouse model of HD has reduced sensitivity to quinolinic acid-induced striatal excitoxicity[134].

11. CONCLUSIONS

There is currently substantial evidence for mitochondrial dysfunction in a wide range of neurodegenerative diseases and the ageing brain. Future studies will need to dissect out the molecular mechanisms involved in causing defects of oxidative phosphorylation and other mitochondrial functions. How this process leads to cell death over many years is still conjectural. The understanding of the interrelationship of apoptosis, excitotoxicity, free radical formation and respiratory chain dysfunction will be critical in answering this question and in finding targets for therapeutic intervention in the future.

REFERENCES

1. Gray MW (1992) The endosymbiont hypothesis revisited. Int Rev Cytol 41: 233-357.
2. Galper JB, Darnell JE (1969) The presence of N-formyl-methionyl-tRNA in HeLa cell mitochondria. Biochem Biophys Res Commun 34: 205-214.
3. Epler JL, Shugart LR, Barnett WE (1970) N-formylmethionyl transfer ribonucleic acid in mitochondria from Neurospora. Biochemistry 9: 3575-3579.
4. Beal MF, Howell N, Bodis-Wollner I (1997) Mitochondria and free radicals in neurodegenerative diseases. Wiley-Liss, New York.
5. Anderson S, Bankier AT, Barrell BG et al. (1981) Sequence and organisation of the human mitochondrial genome. Nature 290: 457-465.
6. Taanman JW (1999) The human mitochondrial genome: structure, transcription, translation and replication. Biochim Biophys Acta 1410: 103-123.
7. Chinnery F, Howell N, Andrews RA (1999) Clinical mitochondrial genetics. J Med Genet 36: 425-436.
8. Lightowlers RN, Chinnery PF, Turnbull DM et al. (1997) Mammalian mitochondrial genetics: heredity, heteroplasmy and disease. Trends Genet 13: 450-455.
9. Leonard JV, Schapira AHV (2000) Mitochondrial respiratory chain disorders I: mitochondrial DNA defects. Lancet 355: 299-304.
10. Leonard JV, Schapira AHV (2000) Mitochondrial respiratory chain disorders II: neurodegenerative disorders and nuclear gene defects. Lancet 355: 389-394.
11. Schapira AHV, Cooper JM, Dexter D et al. (1989) Mitochondrial complex I deficiency in Parkinson's disease. Lancet 1: 1269.
12. Schapira AHV, Mann VM, Cooper JM et al. (1990) Anatomic and disease specificity of NADH CoQ1 reductase (complex I) deficiency in Parkinson's disease. J Neurochem 55: 2142-2145.
13. Gu M, Gash MT, Cooper JM et al. (1997) Mitochondrial respiratory chain function in multiple system atrophy. Mov Disord 12: 418-422.
14. Gu M, Owen AD, Toffa SEK et al. (1998) Mitochondrial function, GSH and iron in neurodegeneration and Lewy body diseases. J Neurol Sci 158: 24-29.
15. DiMauro S (1993) Mitochondrial involvement in Parkinson's disease: the controversy continues. Neurology 43: 2170-2172.
16. Taylor DJ, Krige D, Barnes PRJ et al. (1994) A ^{31}P magnetic resonance spectroscopy study of mitochondrial function in skeletal muscle of patients with Parkinson's disease. J Neurol Sci 125: 77-81.
17. Penn AMW, Roberts T, Hodder J et al. (1995) Generalised mitochondrial dysfunction in Parkinson's disease detected by magnetic resonance spectroscopy of muscle. Neurology 45: 2097-2099.
18. Schapira AHV (1994) Evidence for mitochondrial dysfunction in Parkinson's disease - a critical appraisal. Mov Disord 9: 125-138.
19. Gu M, Cooper JM, Taanman JW et al. (1998) Mitochondrial DNA transmission of the mitochondrial defect of Parkinson's disease. Ann Neurol 44: 177-186.
20. Swerdlow RH, Parks JK, Miller SW et al. (1996) Origin and functional consequences of the complex I defect in Parkinson's disease. Ann Neurol 40: 663-671.
21. Davis GC, Williams AC, Markey SP et al. (1979) Chronic parkinsonism secondary to intravenous injection of meperidine analogues. Psychiatry Res 1: 249-254.
22. Vingerhoets FJG, Snow BJ, Tetrud JJ et al. (1994) Positron emission tomographic evidence for progression of human MPTP-induced dopaminergic lesions. Ann Neurol 36: 765-770.

23. Ramsay RR, Krueger MJ, Youngster SK et al. (1991) Interaction of 1-methyl-4-phenylpyridinium ion (MPP^+) and its analogs with rotenone/piericidin binding site of NADH dehydrogenase. J Neurochem 56: 1184-1190.
24. Cleeter MJW, Cooper JM, Schapira AHV (1992) Irreversible inhibition of mitochondrial complex I by 1-methyl-4-phenylpyridinium: evidence for free radical involvement. J Neurochem 58: 786-789.
25. Adams JD, Klaidman LK, Odunze IN (1989) Oxidative effects of MPTP in the midbrain. Res Commun Subst Abuse 10: 169-180.
26. Hasegawa E, Takeshige K, Oishi T et al. (1990) 1-methyl-4-phenylpyridinium (MPP+) induces NADH-dependent superoxide formation and enhances NADH-dependent lipid peroxidation in bovine heart submitochondrial particles. Biochem Biophys Res Commun 170: 1049-1055.
27. Hantraye P, Brouillet E, Ferrante R et al. (1996) Inhibition of neuronal nitric oxide synthase prevents MPTP-induced parkinsonism in baboons. Nature Med 1017-1021.
28. Przedborski S, Jackson-Lewis V, Yokoyama R et al. (1996) Role of neuronal nitric oxide in 1-methyl-4-phenyl-1,2,3,6-tetrahydropyridine (MPTP)-induced dopaminergic neurotoxicity. Proc Natl Acad Sci USA 93: 4565-4571.
29. Di Monte DA, Royland JE, Anderson A et al. (1997) Inhibition of monoamine oxidase contributes to the protective effect of 7-nitroindazole against MPTP neurotoxicity. J Neurochem 69: 1771-1773.
30. Niwa T, Takeda N, Sasaoka T et al. (1989) Detection of tetrahydroisoquinoline in parkinsonian brain as an endogenous amine by use of gas chromatography-mass spectrometry. J Chromatogr 491: 397-403.
31. Niwa T, Takeda N, Kaneda N et al. (1987) Presence of tetrahydroisoquinoline and 2-methyl-tetrahydroquinoline in parkinsonian and normal human brains. Biochem Biophys Res Commun 144: 1084-1089.
32. Suzuki K, Mizuno Y, Yoshida M (1989) Selective inhibition of complex I of the brain electron transport system by tetrahydroisoquinoline. Biochem Biophys Res Commun 162: 1541-1545.
33. The Huntington's Disease Collaborative Research Group (HDCRG) (1993) A novel gene containing a trinucleotide repeat that is expanded and unstable on Huntington's disease chromosomes. Cell 72: 971-983.
34. Vonsattel JP, Myers RH, Stevens TJ et al. (1985) Neuropathological classification of Huntington's disease. J Neuropath Exp Neurol 44: 559-577.
35. Jenkins BG, Koroshetz WJ, Beal MF (1993) Evidence for impairment of energy metabolism in vivo in Huntington's disease using localised ^1H NMR spectroscopy. Neurology 43: 2689-2695.
36. Davie CA, Barker GC, Quinn N et al. (1994) Proton MRS in Huntington's disease. Lancet 343: 1580.
37. Martin WRW, Clark C, Ammann W et al. (1992) Cortical glucose metabolism in Huntington's disease. Neurology 42: 223-229.
38. Jenkins BG, Rosas HD, Chen YCI et al. (1998) ^1H NMR spectroscopy studies of Huntington's disease. Correlations with CAG repeat numbers. Neurology 50: 1357-1365.
39. Hoang TQ, Bluml S, Dubowitz DJ et al. (1998) Quantitative proton-decoupled 31P MRS and ^1H MRS in the evaluation of Huntington's and Parkinson's diseases. Neurology 50: 1033-1040.
40. Koroshetz WJ, Jenkins BG, Rosen BR et al. (1997) Energy metabolism defects in Huntington's disease and effects of coenzyme Q_{10}. Ann Neurol 41: 160-165.

41. Gu M, Gash MT, Mann VM et al. (1996) Mitochondrial defect in Huntington's disease caudate nucleus. Ann Neurol 39: 385-389.
42. Browne SE, Bowling AC, MacGarvey U et al. (1997) Oxidative damage and metabolic dysfunction in Huntington's disease: selective vulnerability of the basal ganglia. Ann Neurol 41: 646-653.
43. Tabrizi SJ, Cleeter MWJ, Xuereb J et al. (1999) Biochemical abnormalities and excitotoxicity in Huntington's disease brain. Ann Neurol 45: 25-32.
44. Hausladen A, Fridovich I (1994) Superoxide and peroxynitrite inactivate aconitases, but nitric oxide does not. J Biol Chem 269: 29405-29408.
45. Gardner PR, Nguyen DH, White CW (1994) Aconitase is a sensitive and critical target of oxygen poisoning in cultured mammalian cells and in rat lungs. Proc Natl Acad Sci USA 91: 12248-12252.
46. Patel M, Day BJ, Crapo JD et al. (1996) Requirement for superoxide in excitotoxic cell death. Neuron 16: 345-355.
47. Parker WD, Boyson SJ, Luder AS et al. (1990). Evidence for a defect in NADH: ubiquinone oxidoreductase (complex I) in Huntington's disease. Neurology 40: 1231-1234.
48. Arenas J, Campos Y, Ribacoba R et al. (1998) Complex I defect in muscle from patients with Huntington's disease. Ann Neurol 43: 397-400.
49. Schapira AHV (1997) Mitochondrial function in Huntington's disease: clues for the pathogenesis and prospects for treatment. Ann Neurol 41: 141-142.
50. Schapira AHV (1999) Mitochondrial involvement in Parkinson's disease, Huntington's disease, hereditary spastic paraplegia, and Friedreich's ataxia. Biochim Biophys Acta 1210: 159-170.
51. Beal MF, Brouillet E, Jenkins BG et al. (1993) Neurochemical and histological characterisation of striatal excitotoxic lesions produced by the mitochondrial toxin 3-nitropropionic acid. J Neurosci 13: 4181-4192.
52. Mangiarini L, Sathasivam K, Seller M et al. (1996) Exon 1 of the HD gene with an expanded CAG repeat is sufficient to cause a progressive neurological phenotype in transgenic mice. Cell 87: 493-506.
53. Tabrizi SJ, Workman J. Hart P et al. (2000) Mitochondrial dysfunction and free radical damage in the Huntington R6/2 transgenic mouse. Ann Neurol 47: 80-86.
54. Cha JHJ, Kosinski CM, Kerner JA et al. (1998) Altered brain neurotransmitter receptors in transgenic mice expressing a portion of an abnormal human Huntington disease gene. Proc Natl Acad Sci USA 95: 6480-6485.
55. Levine MS, Chesselet MF, Koppel A et al. (1998) Enhanced sensitivity to glutamate receptor activation in mouse models of Huntington's disease. J Neurosci 24: 972.
56. Bogdanov M, Ferrante RJ, Kuemmerle S et al. (1998) Increased vulnerability to 3-nitropropionic acid in an animal model of Huntington's disease. J Neurochem 71: 2642-2644.
57. De Michele G, De Fusco M, Cavalcanti F et al. (1998) A new locus for autosomal recessive hereditary spastic paraplegia. Am J Hum Genet 63: 135-139.
58. Campuzano V, Montermini L, Molto MD et al. (1996) Friedreich's ataxia: autosomal recessive disease caused by an intronic GAA triplet repeat expansion. Science 217: 1423-1427.
59. Campuzano V, Montermini L, Lutz Y et al. (1997) Frataxin is reduced in Friedreich ataxia patients and is associated with mitochondrial membranes. Hum Mol Genet 6: 1771-1780.

60. Koutnikova H, Campuzano V, Foury F et al. (1997) Studies of human, mouse and yeast homologues indicate a mitochondrial function for frataxin. Nature Genet 16: 345-351.
61. Priller J, Scherzer CR, Faber PW et al. (1997) Frataxin gene of Friedreich's ataxia is targeted to mitochondria. Ann Neurol 42: 265-269.
62. Wilson RB, Roof DM (1997) Respiratory deficiency due to a loss of mitochondrial DNA in yeast lacking frataxin homologue. Nature Genet 16: 352-357.
63. Sanyal A, Harrington A, Herbert CJ et al. (1995) Heat shock protein HSP60 can alleviate the phenotype of mitochondrial RNA-deficient temperature-sensitive mna2 pet mutants. Mol Gen Genet 246: 56-64.
64. Sanchez-Casis G, Cote M, Barbeau A (1976) Pathology of the heart in Friedreich's ataxia. Can J Neurol Sci 3: 349-354.
65. Thomas PK, Cooper JM, King RHM et al. (1993) Myopathy in vitamin E deficient rats: muscle fibre necrosis associated with disturbances of mitochondrial function. J Anat 183: 451-461.
66. Finocchiaro G, Baio G, Micossi P et al. (1988) Glucose metabolism alterations in Friedreich's ataxia. Neurology 38: 1292-1296.
67. Rotig A, De Lonlay P, Chretien D et al. (1997) Aconitase and mitochondrial iron-sulphur protein deficiency in Friedreich's ataxia. Nature Genet 17: 215-217.
68. Bradley JL, Blake J, Chamberlain S et al. (2000) Clinical, biochemical and molecular genetic correlations in Friedreich's ataxia. Hum Mol Genet 9: 275-282.
69. Lodi R, Cooper JM, Bradley JL et al. (1999) Deficit of in vivo mitochondrial ATP production in patients with Friedreich's ataxia. Proc Natl Acad Sci USA 96: 11492-11495.
70. Katzman R (1986) Alzheimer's disease. New Engl J Med 314: 457-465.
71. Duara R, Lopez-Alberola RF, Barker WW et al. (1993) A comparison of familial and sporadic Alzheimer's disease. Neurology 43: 1377-1384.
72. Mielke R, Herholz K, Grond M et al. (1994) Clinical deterioration in probable Alzheimer's disease correlates with progressive metabolic impairment of association areas. Dementia 5: 36-41.
73. Friedland RP, Budinger TF, Koss E et al. (1985) Alzheimer's disease: anterior-posterior and lateral hemispheric alterations in cortical glucose utilization. Neurosci Lett 33: 235-240.
74. Parker WD, Parks JK, Filley CM et al. (1994) Electron transport chain defects in Alzheimer's disease brain. Neurology 44: 1090-1096.
75. Parker WD, Parks JK (1995) Cytochrome c oxidase in Alzheimer's disease brain: Neurology 45: 482-486.
76. Kish SJ, Bergeran C, Rajput A et al. (1992) Brain cytochrome oxidase in Alzheimer's disease. J Neurochem 59: 776-779.
77. Mutisya EM, Bowling AC, Beal MF (1994) Cortical cytochrome oxidase activity is reduced in Alzheimer's disease. J Neurochem 63: 2179-2184.
78. Chagnon P, Betard C, Robitaille Y et al. (1995) Distribution of brain cytochrome oxidase activity in various neurodegenerative diseases. Mol Neurosci 6: 711-715.
79. Simonian NA, Hyman BT (1993) Functional alterations in Alzheimer's disease: diminution of cytochrome oxidase in the hippocampal formation. J Neuropathol Exp Neurol 52: 580-585.
80. Simonian NA, Hyman BT (1994) Functional alterations in Alzheimer's disease: selective loss of mitochondrial-encoded cytochrome oxidase mRNA in hippocampal formation. J Neuropathol Exp Neurol 53: 508-512.

81. Bonilla E, Tanji K, Hirano M et al. (1999) Mitochondrial involvement in Alzheimer's disease. Biochim Biophys Acta 1410: 171-182.
82. Hevner RF, Wong-Riley MTT (1993) Mitochondrial and nuclear gene expression for cytochrome oxidase subunits are disproportionately regulated by functional activity in neurons. J Neurosci 13: 1805-1819.
83. Corral-Debrinski M, Horton T, Lott MT et al. (1994) Marked changes in mitochondrial DNA deletion levels in Alzheimer brains. Genomics 23: 471-476.
84. Blanchard BJ, Park T, Fripp WJ et al. (1993) A mitochondrial DNA deletion in normally aging and Alzheimer brain tissue. Neuroreport 4: 799-802.
85. Chandrasekaran K, Hatanpaa K, Brady DR et al. (1996) Evidence for physiological down-regulation of brain oxidative phosphorylation in Alzheimer's disease. Exp Neurol 142: 80-88.
86. Edland SD, Silverman JM, Peskind ER et al. (1996) Increased risk of dementia in mothers of Alzheimer's disease cases:evidence for maternal inheritance. Neurology 47: 254-256.
87. Shoffner JM, Brown MD, Torroni A et al. (1993) Mitochondrial DNA variants observed in Alzheimer disease and Parkinson's disease patients. Genomics 17: 171-184.
88. Wragg MA, Talbot CJ, Morris JC et al. (1995) No association found between Alzheimer's disease and a mitochondrial tRNA glutamine gene variant. Neurosci Lett 201: 107-110.
89. Hutchin T, Cortopassi G (1995) A mitochondrial DNA clone is associated with increased risk for Alzheimer disease. Proc Natl Acad Sci USA 92: 6892-6895.
90. Davis RE, Miller S, Herrnstadt C et al. (1997) Mutations in mitochondrial cytochrome c oxidase genes segregate with late-onset Alzheimer disease. Proc Natl Acad Sci USA 94: 4526-4531.
91. Swerdlow RH, Parks JK, Cassarino DS et al. (1997) Cybrids in Alzheimer's disease: a cellular model of disease? Neurology 49: 918-925.
92. Schon EA, Shoubridge EA, Moraes CT (1998) Cybrids in Alzheimer's disease: a cellular model of disease? Neurology 51: 326-327.
93. Coyle JT, Puttfarcken P (1993) Oxidative stress, glutamate and neurodegenerative disorders. Science 262: 689-695.
94. Hensley K, Hall N, Subramanian R et al. (1995) Brain regional correspondence between Alzheimer's disease histopathology and biomarkers of protein oxidation. J Neurochem 65: 2146-2156.
95. Lovell MA, Ehman WD, Butler SM et al. (1995) Elevated thiobarbituric acid-reactive substances and antioxidant enzyme activity in the brain in Alzheimer's disease. Neurology 45: 1594-1601.
96. Good PF, Werner P, Hsu A et al. (1996) Evidence for neuronal oxidative damage in Alzheimer's disease. Am J Pathol 149: 21-28.
97. Smith MA, Perry G, Richey PL et al. (1996) Oxidative damage in Alzheimer's. Nature 382: 120-121.
98. Ledesma MD, Boney P, Colaco C et al. (1994) Analysis of microtubule-associated protein tau glycation in paired helical filaments. J Biol Chem 269: 21614-21619.
99. Vitek MP, Bhattacharya K, Glendening JM (1994) Advanced glycation end-products contribute to amyloidosis in Alzheimer's disease. Proc Natl Acad Sci USA 91: 4766-4770.

100. Gsell W, Conrad R, Hickethier M et al. (1995) Decreased catalase activity but unchanged superoxide dismutase activity in brains of patients with dementia of Alzheimer type. J Neurochem 64: 1216-1223.
101. Griggs RC, Askanas V, DiMauro S et al. (1995) Inclusion body myositis and myopathies. Ann Neurol 38: 705-713.
102. Mecocci P, MacGarvey U, Kaufman AE et al. (1993) Oxidative damage to mitochondrial DNA shows marked age-dependent increases in human brain. Ann Neurol 34: 609-616.
103. Mecocci P, MacGarvey U, Beal MF (1994) Oxidative damage to mitochondrial DNA is increased in Alzheimer's disease. Ann Neurol 36: 747-751.
104. Morley AA (1995) The somatic mutation theory of ageing. Mutat Res 338: 19-23.
105. Brennan JA, Boyle JO, Koch WM et al. (1995) Association between cigarette smoking and mutation of the p53 gene in squamous cell carcinoma of the head and neck. New Engl J Med 332: 712-717.
106. Harman D (1972) The biologic clock: the mitochondria? J Am Ger Soc 20: 145-147.
107. Corral-Debrinski M, Horton T, Lott MT et al. (1992) Mitochondrial DNA deletions in human brain: regional variability and increase with advanced age. Nature Genet 2: 324-329.
108. Wallace DC (1992) Mitochondrial genetics: a paradigm for aging and degenerative disease? Science 256: 628-632.
109. Cortopassi GA, Shibata D, Soong NW et al. (1992) A pattern of accumulation of a somatic deletion of mitochondrial DNA in aging human tissues. Proc Natl Acad Sci USA 89: 7370-7374.
110. Soong NW, Hinton DR, Cortopassi G et al. (1992) Mosaicism for a specific somatic mitochondrial DNA mutation in adult human brain. Nature Genet 2: 1075-1078.
111. Holt I.J, Harding AE, Morgan-Hughes JA (1988) Deletions of muscle mitochondrial DNA in patients with mitochondrial myopathies. Nature 331: 717-719.
112. Schon EA, Rizzuto R, Moraes CT et al. (1989) A direct repeat is a hotspot for large-scale deletions of human mitochondrial DNA. Science 244: 346-349.
113. Mita S, Rizzuto R, Moraes CT et al. (1990) Recombination via flanking direct repeats is a major cause of large-scale deletions of human mitochondrial DNA. Nucleic Acids Res 18: 561-567.
114. Nakase H, Moraes CT, Rizzuto R et al. (1990) Transcription and translation of deleted mitochondrial genomes in Kearns-Sayre syndrome:implications for pathogenesis. Am J Hum Genet 46: 418-427.
115. Simonetti S, Chen X, DiMauro S et al. (1992) Accumulation of deletions in human mitochondrial DNA during normal aging: analysis by quantitative PCR. Biochim Biophys Acta 1180: 113-122.
116. Lee CM, Weindruch R, Aiken JM (1997) Age-associated alterations of the mitochondrial genome. Free Rad Biol Med 22: 1259-1269.
117. Liu VW, Zhang C, Linnane AW et al. (1997) Quantitative allele-specific PCR: demonstration of age-associated accumulation in human tissues of the A→G mutation at nucleotide 3243 in mitochondrial DNA. Hum Mutat 9: 265-271.
118. Pallotti F, Chen X, Bonilla E et al. (1996) Evidence that specific mtDNA point mutations may not accumulate in skeletal muscle during normal human aging. Am J Hum Genet 59: 591-602.
119. Kadenbach B, Munscher C, Frank V et al. (1995) Human aging is associated with stochastic somatic mutations of mitochondrial DNA. Mutat Res 338: 161-172.

120. Brierly EJ, Johnson MA, James OF et al. (1996) Effects of physical activity and age on mitochondrial function. Q J Med 89: 251-258.
121. Papa S (1996) Mitochondrial oxidative phosphorylation changes in the life span. Molecular aspects and physiopathological implications. Biochim Biophy Acta 1276: 87-105.
122. Barrientos A, Casademont J, Rotig A et al. (1996) Absence of relationship between the level of electron transport chain activities and aging in human skeletal muscle. Biochem Biophys Res Commun 229: 536-539.
123. Linnane AW, Degli Esposti M, Generowicz M et al. (1995) The universality of bioenergetic disease and amelioration with redox therapy. Biochim Biophys Acta 1271: 191-194.
124. Hagen TM, Yowe DL, Bartholomew JC et al. (1997) Mitochondrial decay in hepatocytes from old rats: membrane potential declines, heterogeneity and oxidants increase. Proc Natl Acad Sci USA 94: 3064-3069.
125. Tatton WG, Olanow CW (1999) Apoptosis in neurodegenerative diseases: the role of mitochondria. Biochim Biophys Acta 1410; 195-213.
126. Leist M, Single B, Castoldi AF et al. (1997) Intracellular adenosine triphosphate (ATP) concentration: a switch in the decision between apoptosis and necrosis. J Exp Med 185: 1481-1486.
127. Susin SA, Zamzami N, Kroemer G (1998) Mitochondria as regulators of apoptosis: doubt no more. Biochim Biophys Acta 1366: 151-156.
128. Shimizu S, Narita M, Tsujimoto Y (1999) Bcl-2 family proteins regulate the release of apoptogenic cytochrome c by the mitochondrial channel VDAC. Nature 1366: 139-149.
129. Liu P, Kim CN, Yang J et al. (1996) Induction of apoptotic program in cell-free extracts: requirement for dATP and cytochrome c. Cell 86: 147-157.
130. Li P, Nijhawan D, Budihardo I et al. (1997) Cytochrome c and dATP-dependent formation of Apaf-1/caspase-9 complex initiates an apoptotic protease cascade. Cell 91: 479-489.
131. Susin SA, Zamzami N, Castedo M et al. (1996) Bcl-2 inhibits the mitochondrial release of an apoptotic protease. J Exp Med 184: 1331-1341.
132. Choi DW (1996) Ischemia-induced neuronal apoptosis. Curr Opin Neurobiol 6: 667-672.
133. Bernardi P, Petronilli V (1996) The permeability transition pore as a mitochondrial calcium release channel: a critical appraisal. J Bioenerg Biomembr 28: 131-138.
134. Hansson O, Petersen A, Leist M et al. (1999) Transgenic mice expressing a Huntington's disease mutation are resistant to quinolinic acid-induced striatal excitotoxicity. Proc Natl Acad Sci USA 96: 8727-8732.

Index

GPSR Compliance

The European Union's (EU) General Product Safety Regulation (GPSR) is a set of rules that requires consumer products to be safe and our obligations to ensure this.

If you have any concerns about our products, you can contact us on ProductSafety@springernature.com

In case Publisher is established outside the EU, the EU authorized representative is:

Springer Nature Customer Service Center GmbH
Europaplatz 3
69115 Heidelberg, Germany

Batch number: 10755809

Printed by Printforce, the Netherlands